W9-BHL-813
3 1192 00772 5047

EVERYTHING YOU NEED TO KNOW ABOUT

HOMEWORK

Anne Zeman and Kate Kelly

An Irving Place Press Book

Scholastic Reference

New York Toronto London Auckland Sydney

Design: Bennett Gewirtz, Gewirtz Graphics, Inc.
Illustration: Moffitt Cecil

For their assistance in the preparation of this manuscript, grateful acknowledgment to Betty Holmes, Director of UFT's Dial-A-Teacher, Vincent Ridge, and Carmen Edgerly. Dial-A-Teacher is a collaborative program of the United Federation of Teachers and the New York City Board of Education. Some of the illustrations in this book were previously published in the Scholastic Environmental Atlas of the United States by Mark Mattson, Copyright © 1993 by Scholastic Inc.

Grateful acknowledgment is made to:
Steven W. Jones, FPG International, for permission to reprint photograph on page 14.
Gregory G. Dimijian, Photo Researchers, Inc., for permission to reprint photograph on page 33.
Ralph A. Reinhold, FPG International, for permission to reprint photograph on page 38.
Joyce Photographics for photograph of igneous rock, Rod Planck for photograph of sandstone, and Adam Hart-Davis, Science Photo Library, for photograph of marble, all from Photo Researchers, Inc., for permission to reprint photographs on page 48.
NASA, FPG International, for permission to reprint photographs of Mercury, Venus, and Uranus on pages 86 and 87.
NASA for permission to reprint photographs on pages 83, 86, 87, 89, 90, and 91.
Telegraph Colour Library, FPG International, for permission to reprint photograph on page 101.

Library of Congress Cataloging-in-Publication Data

Zeman, Anne, 1951-
Everything you need to know about science homework / Anne Zeman and Kate Kelly.
p. cm. — (Scholastic homework reference series)
Includes index.
ISBN 0-590-49356-6
1. Science—Study and teaching—Handbooks, manuals, etc.—Juvenile literature.
2. Homework—Handbooks, manuals, etc.—Juvenile literature.
[1. Science. 2. Homework.] I. Kelly, Kate. II. Title. III. Series.
Q181.Z45 1994
500—dc20 93-49352
CIP
AC

12 11 10 9 8 7 6 5 4 3 2 1 4 5 6 7 8 9/9

Printed in the U.S.A.

First Scholastic printing, August 1994

Contents

Part 5. Planet Earth

Part 6. Outer Space

Part 7. The Physical World

INTRODUCTION

It's homework time—but you have questions. Just how did your teacher ask you to do the assignment? You need help, but your parents are busy and you can't reach your classmate on the phone. Where can you go for help?

What Questions Does This Book Answer?

In *Everything You Need to Know About Science Homework*, you will find a wealth of information, including the answers to ten of the most commonly asked science homework questions.*

1. What is the difference between a plant cell and an animal cell? Plant and animal cells are illustrated on page 6.
2. Are there any mammals that lay eggs? Mammals are described on page 13.
3. How does water move up and down a plant? You'll find an illustration on page 23.
4. What is the hardness scale for rocks? The Mohs scale is described in the text and on a chart on page 50.
5. What is the difference between a breeze, a gale, a storm, and a hurricane? The Beaufort Scale on page 74 has the answer.
6. What are the phases of the moon? The moon's phases are illustrated on page 91.
7. What are the three states of matter? States of matter are defined on page 94.
8. Why do some things sink and others float? The explanation of density on page 96 provides the answer.
9. How do you measure the strength of an acid? A description of acids, bases, and litmus paper is located on page 100.
10. What is color? What are black and white? Colors, and black and white, are described on page 116.

Note: You won't find answers to homework questions on the human body in this book. Because the subject is so important and complex, it is the subject of its own Homework Reference volume.

* According to Dial-A-Teacher

What Is the Scholastic Homework Reference Series?

The Scholastic Homework Reference Series is a set of unique reference resources written especially to answer the homework questions of fourth, fifth, and sixth graders. The series provides ready information to answer commonly asked homework questions in a variety of subjects. Here you'll find facts, charts, definitions, and explanations, complete with examples and illustrations that will supplement schoolwork colorfully, clearly—and comprehensively.

A Note to Parents

The information for the Scholastic Homework Reference Series was gathered from current textbooks, national curricula, and the invaluable assistance of the UFT Dial-A-Teacher staff. Dial-A-Teacher, a collaborative program of the United Federation of Teachers and the New York City Board of Education, is a telephone service available to elementary school students in New York City. Telephone lines are open during the school term from 4:00 to 7:00 p.m., Monday to Thursday, by dialing 212-777-3380. Because of Dial-A-Teacher's success in New York City, similar organizations have been established in other communities across the country. Check to see if there's a telephone homework service in your area.

It's important to support your children's efforts to do homework. Welcome their questions and see that they are equipped with a well-lighted desk or table, pencils, paper, and any other books or equipment—such as rulers, calculators, reference or text books, and so on—that they may need. You might also set aside a special time each day for doing homework, a time when you're available to answer questions that may arise. But don't do your children's homework for them. Remember, homework should create a bond between school and home. It is meant to enhance on a daily basis the lessons taught at school, and to promote good work and study habits. Although it is gratifying to have your children present flawless homework papers, the flawlessness should be a result of your children's explorations and efforts—not your own.

The Scholastic Homework Reference Series is designed to help your children complete their homework on their own to the best of their abilities. If they're stuck, you can use these books with them to find answers to troubling homework problems. And, remember, when the work is done—praise your children for a job well done.

EVERYTHING YOU NEED TO KNOW ABOUT

HOMEWORK

What's Life All About?

1 Living or Nonliving?

Everything on the earth is ***living*** or ***nonliving***. Plants, animals, bacteria, and fungi are all living things. Rocks, clouds, soil, lake water, and metals are nonliving things.

Most living things are easy to tell apart from nonliving things. For example, dogs, cats, flowers, trees, and humans grow and change over time. But other living things, such as bacteria, molds, and mildews, often don't seem to be alive. And some are so tiny that you can see them only with a powerful microscope. Yet all living things—or ***organisms***—can do at least one of the six things that separate living from nonliving things.

Organisms

1. Take in food to provide energy (see Energy, p.101)
2. Grow and develop
3. Give off wastes
4. Respond to their surroundings
5. Reproduce themselves
6. Are made up of cells

The Scientists of Living Things

The science of living things is called ***biology***, and a scientist who studies living things is known as a ***biologist***. Biologists study all kinds of life, from plants and animals to microorganisms, ecology, extinct or endangered species, and more. Here is what some biologists study:

Botanists specialize in ***botany***, the study of plants.

Zoologists specialize in ***zoology***, the study of animals.

Microbiologists specialize in ***microbiology***, the study of ***microscopic*** (super tiny) plants and animals.

Cytologists specialize in ***cytology***, the study of cells.

Ecologists specialize in ***ecology***, the study of the interrelationship of organisms and their environment.

Entomologists specialize in ***entomology***, the study of insects.

Herpetologists specialize in ***herpetology***, the study of reptiles and amphibians.

Ichthyologists specialize in ***ichthyology***, the study of fish.

Mammalogists specialize in ***mammalogy***, the study of mammals.

Marine biologists specialize in ***marine biology***, the study of plants and animals in seas and oceans.

Ornithologists specialize in ***ornithology***, the study of birds.

2 The Five Kingdoms

For scientific study, living things are often placed into five different groups called ***kingdoms***. These five kingdoms include all ***organisms*** (living things), from elephants and trees, to ***microorganisms*** (or microbes), the tiny, usually one-celled organisms that cannot be seen without a microscope.

The Five Kingdoms

Animalia (Animals)

Many-celled organisms that use their own ***nervous system*** to move (see p. 11). Animals cannot make their own food. They must eat other plants and animals to produce energy. Animal cells do not have rigid walls (see p. 6). This kingdom includes mammals, birds, reptiles, amphibians, fish, mollusks, sponges, worms, insects, and spiders.

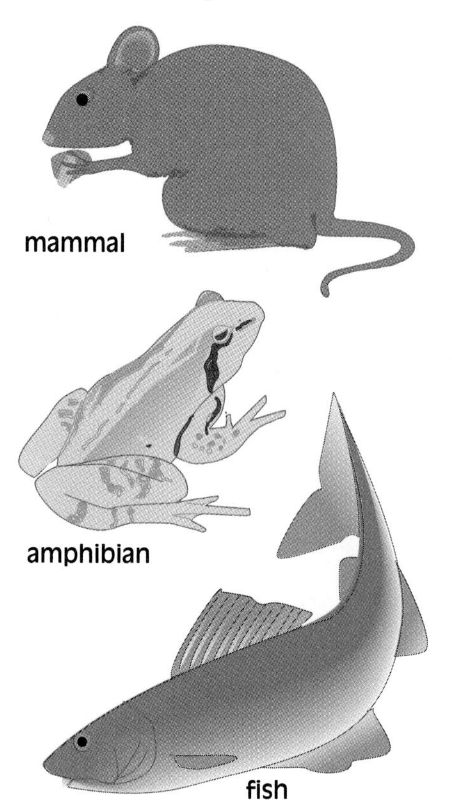

Plantae (Plants)

Many-celled green organisms that make their own food through ***photosynthesis*** (see p. 28). Plant cells have rigid walls with cellulose in them (see p. 6). This kingdom includes flowering plants, trees, mosses, and ferns (see pages 20-21). Some experts include green algae in the plant kingdom. Others put all the algae in the ***Protista*** kingdom.

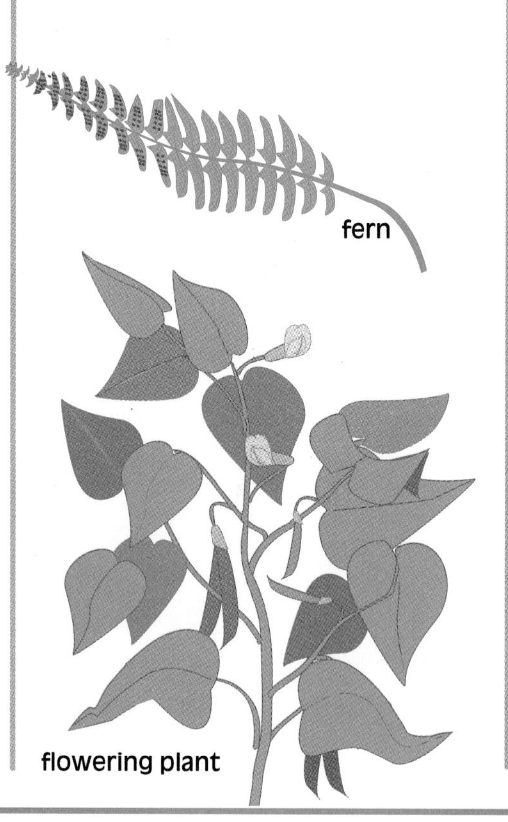

Fungi (Fungus Organism

People once thought that mushrooms and other fungi were pla But fungi have no ***chlorophyll***, substance plants use to make th own food. Instead, fungi dissolv their food and then absorb it. Their cell walls contain mostly ***chitin***, a substance that is also i insect ***exoskeletons*** (see p. 7) and lobster shells. This kingdom includes molds, mildews, rusts, yeasts, mushrooms, penicillin, and mor

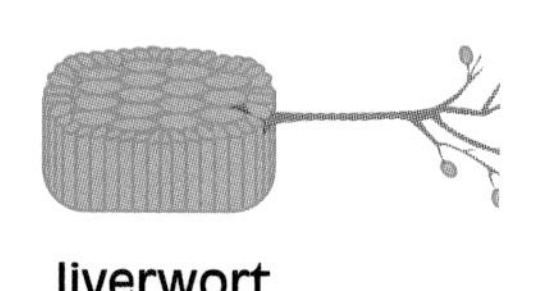

Protista (Protists)

Mostly one-celled micro-organisms, including plant- and animal-like organisms. This kingdom includes amoebas, paramecia, diatoms, euglenas, and many more. All the Protista have cells with a true ***nucleus*** (see p. 6), as do animals, plants, and fungi.

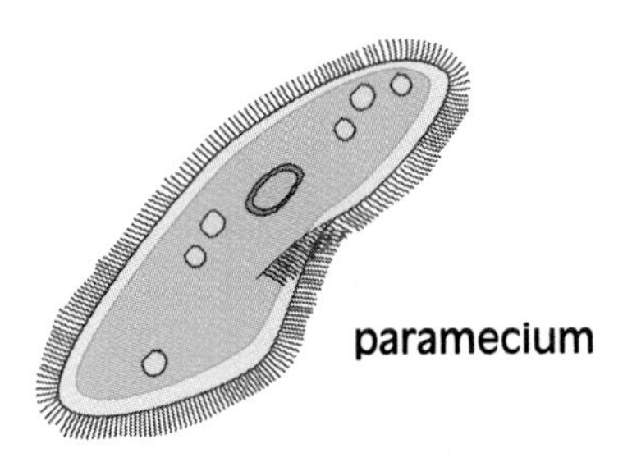

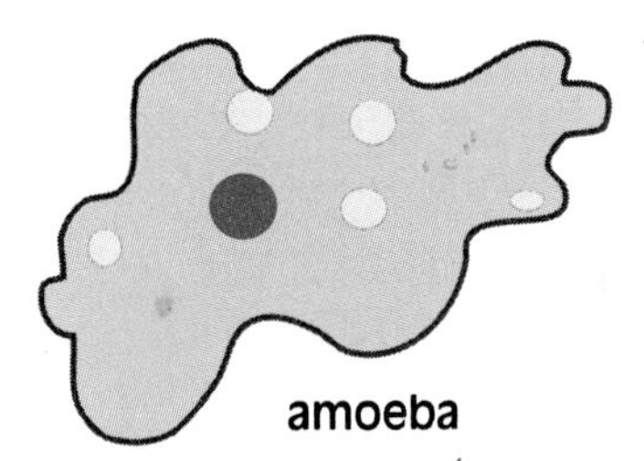

Monera (Monerans)

Microscopic one-celled organisms. Their cells do not have a true nucleus (see p. 6). Most Monera absorb their food. There are thousands of different species of Monera, including bacteria and blue-green algae.

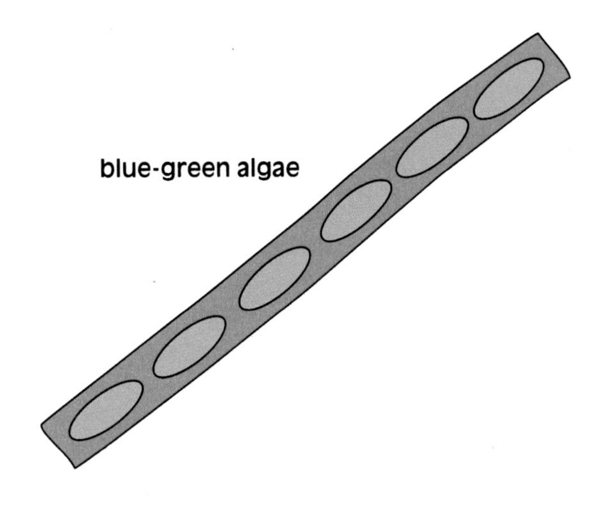

Plant and Animal Cells

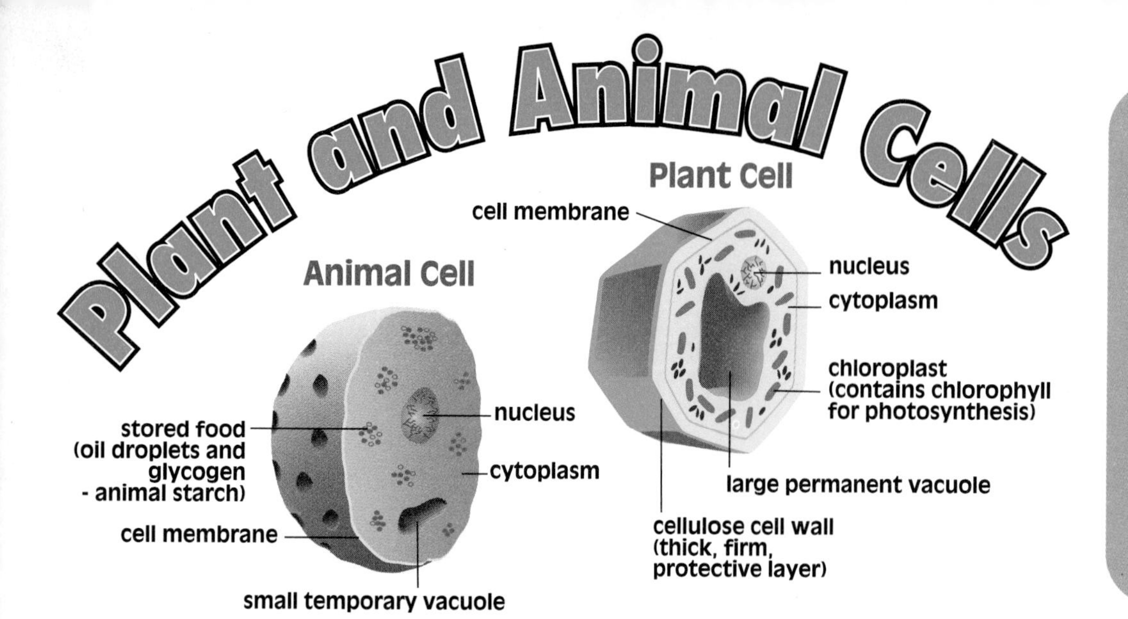

Cells are tiny living things that make up all the plants and animals on the earth. Some plants and animals are made up of only one cell. But the plants and animals we can see around us are made up of millions and millions of cells. That means cells are the tiny building blocks of life.

The King's Subjects: Classifying Living Things

Genus and Species

Carolus Linnaeus (1707-1778), a Swedish ***botanist*** (see p. 3), made up a simple system that scientists still use to name living things. Linnaeus used two Latin words—***genus*** (from the Latin for "race" or "birth") and ***species*** (meaning "kind" of organism). For example, dogs and wolves belong to the same genus, ***Canis*** (dog). But your pet dog is part of the species ***Canis familiaris*** (domestic dog) and a wolf belongs to the species ***Canis lupus*** (wolf dog).

The System

During Linnaeus's time, scientists recognized only two kingdoms of living things: plants and animals. Since then, they have added three more kingdoms: fungi, protists, and monerans. They have also added four more categories between kingdom and genus. Today, eight categories help people understand how living things are related to each other.

How Scientists Classify You!

Kingdom:	Animal
Phylum:	Chordate (animal with a spinal cord)
Subphylum:	Vertebrate (chordate with a backbone)
Class:	Mammal (vertebrate with hair and one bone in lower jaw, warm-blooded, babies drink mother's milk)
Order:	Primate (mammal whose thumbs and fingers work together, has fingernails, good vision, a large brain, weak sense of smell
Family:	Hominid (primate that walks upright, has a flat face, looks forward, sees color, has hands and feet that perform many tasks)
Genus:	Homo (talks, spends long child-hood learning about the world)
Species:	Homo sapiens ("wise man"—little body hair, high forehead, chin that juts out)

Humans are the only hominids left on the earth. The rest have died out or become extinct.

THE ANIMAL KINGDOM

The Invertebrates

Invertebrates are animals without ***backbones***. Some invertebrates, such as worms and jellyfish, have no protection for their soft, fleshy bodies. Others, such as clams and snails, have shells. Still other invertebrates, such as spiders and beetles, have a hard outer skin called an ***exoskeleton.*** Exoskeletons are much like knights' armor. They protect the soft inner bodies of these invertebrates.

Scientists have identified more than one million species of invertebrates and have divided them into many phyla (see The King's Subjects, p. 6). Many more million species of invertebrates have yet to be named and described.

VERTEBRATE or INVERTEBRATE?

All animals belong to one of two large groups, ***vertebrates*** (animals with backbones) or ***invertebrates*** (animals without backbones). Almost all the world's animals are invertebrates. They include insects, spiders, crustaceans (such as crabs and lobsters), snails, jellyfish, worms, and many, many more. The vertebrates include humans and other mammals, fish, reptiles, birds, and amphibians, such as frogs and salamanders. In all, there are about 45,000 species of vertebrates, while there are many million species of invertebrates.

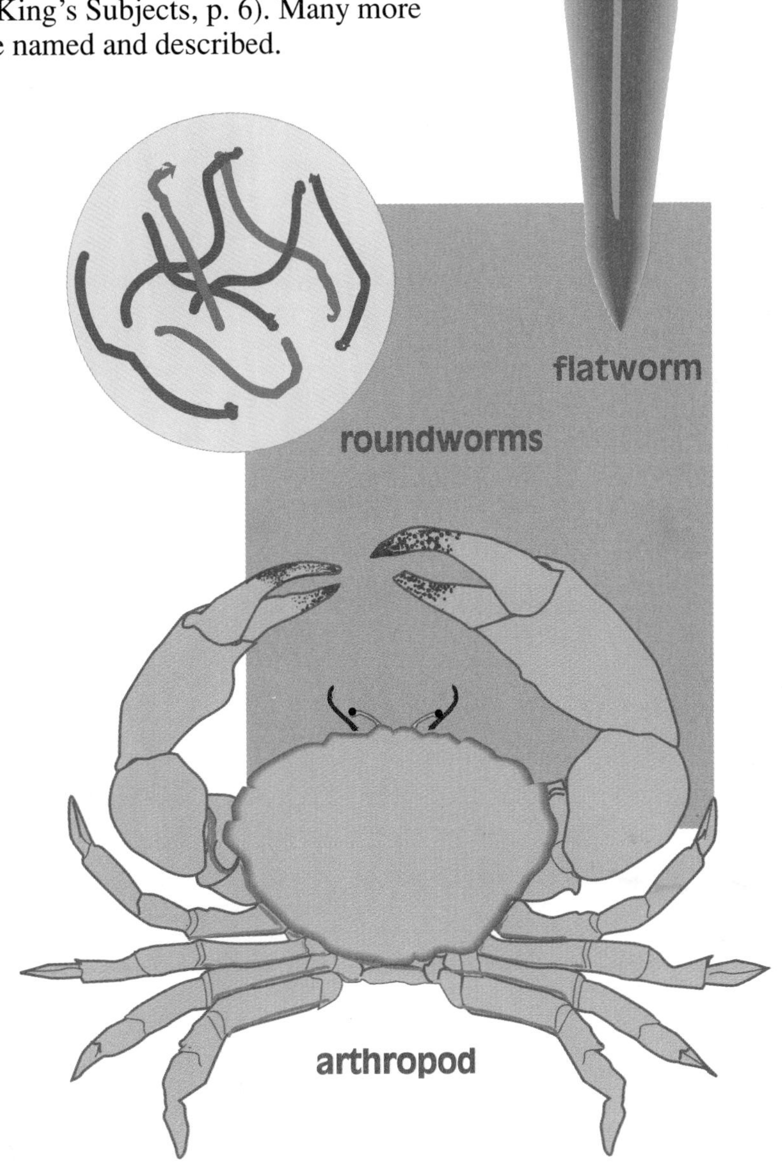

Sponges (Porifera)

The simplest many-celled animal. The sponge is a food-gathering tube. As water passes through the sponge's tubes, special cells trap and digest tiny food particles. Sponges come in many colors, live in water, and usually attach themselves to rocks.

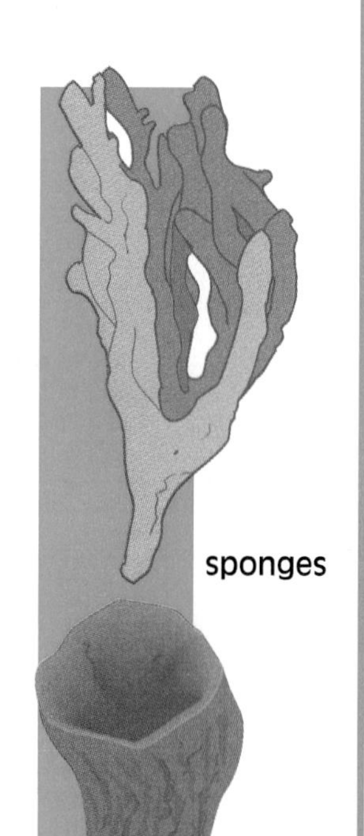

Stinging-cell Animals (Cnidaria)

Animals with a tube or sac-like body with a "mouth" (an opening), which is surrounded by tentacles. Special stinging cells trap prey or keep off ***predators*** (see p.18). Jellyfish, corals, and sea anemones are among the 10,000 or so species of water-dwelling cnidarians.

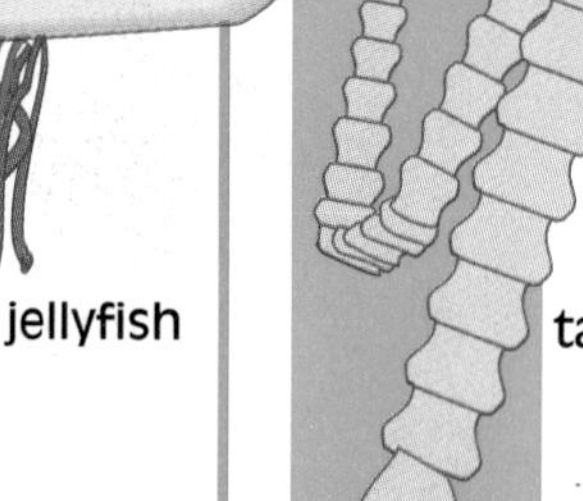

Flatworms (Playthelminthes)

Flat-bodied worms with a head and a simple nervous system. They live in water or moist places. Some are parasites. Among the 13,000 known species of flatworms are tapeworms, flukes, and planaria.

planarian

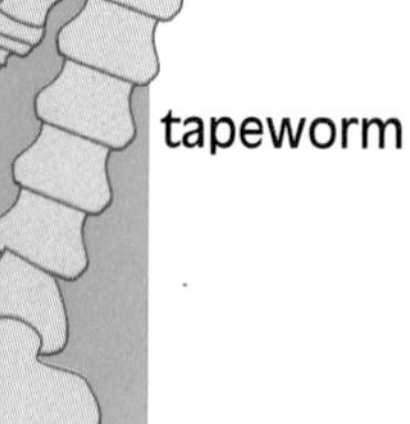

Roundworms (Nematoda)

Smooth, round worms. Many thousands of the 12,000 species of roundworms are microscopic. Some are parasites, and some cause disease.

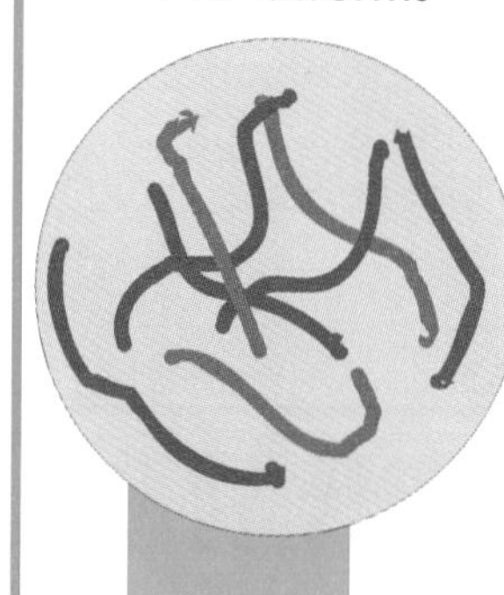

Segmented Worms (Annelida)

Soft body made u of little parts, or ***segments***. Anneli have a true digestive system and tv body openings. Most have well developed circulatory and nervous systems, as well a sense organs.

ollusks (Mollusca)

oft body, usually
nclosed in a hard shell.
ollusk bodies have
ree parts: a head with a
outh, sense organs, and
brain; a body; and a
oot" for creeping
bout. The 50,000
nown species of mol-
sks include snails,
ugs, octopuses, squids,
callops, mussels, and
ysters.

Sea Stars and Their Relatives (Enchinodermata)

Bodies with internal skeletons made mostly of lime. The bodies are arranged in five parts, or in multiples of five. Many of the 5,500 species have body spikes or spines. Sea stars (also called starfish), brittle stars, sand dollars, sea urchins, and sea cucumbers are included in this phylum.

Arthropods (Arthropoda)

The largest phylum, with more than one million known species. All have ***exoskeletons*** made mostly of ***chitin*** (a hard, chalky substance), jointed limbs, and hairs or other bristles that act as sense organs. Almost all are insects, but arthropods (from Greek for "jointed foot") also include spiders, crustaceans such as lobsters and crabs, millipedes, and centipedes.

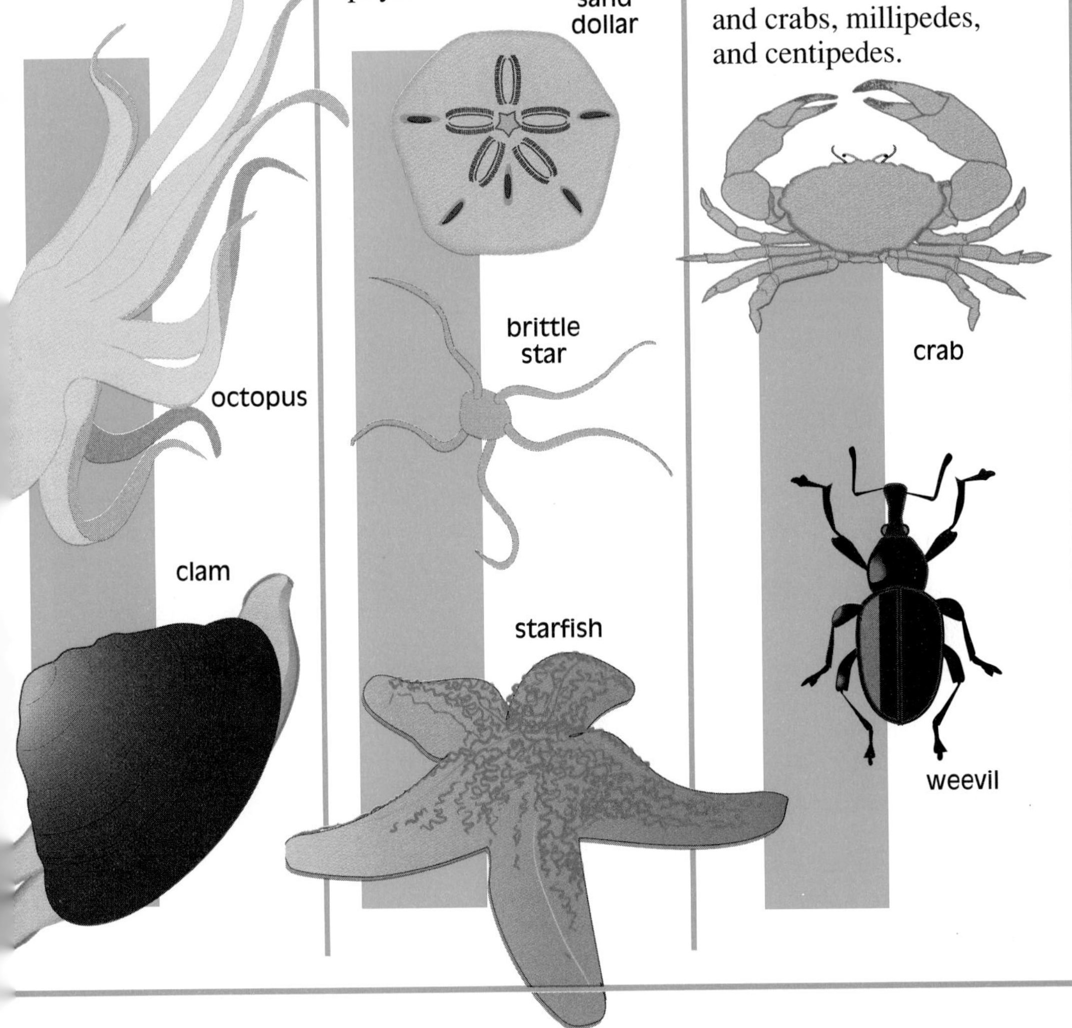

Note: Only invertebrate phyla with more than 5,000 species are included on this chart.

Don't Bug Me!

Some insects are big, some small, some common, some rare. Some insects have wings, some don't. In fact, insects are the only invertebrates that can fly!

Insects have bodies made up of three distinct parts—head, thorax, and abdomen. The eyes, mouth, and antennae are located on the head. The legs and wings are attached to the thorax! The digestive system and reproductive organs are located in the abdomen.

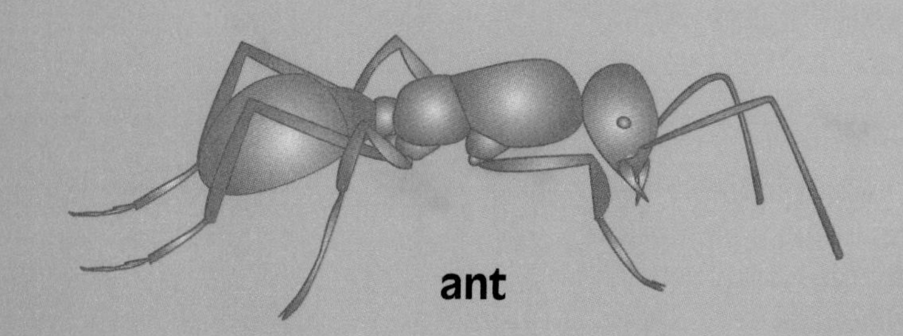

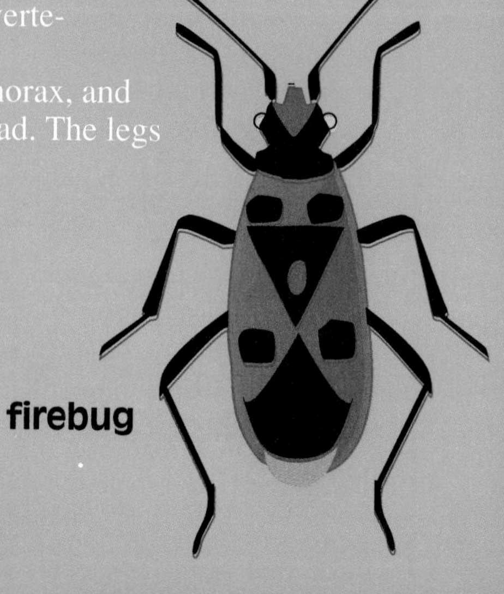

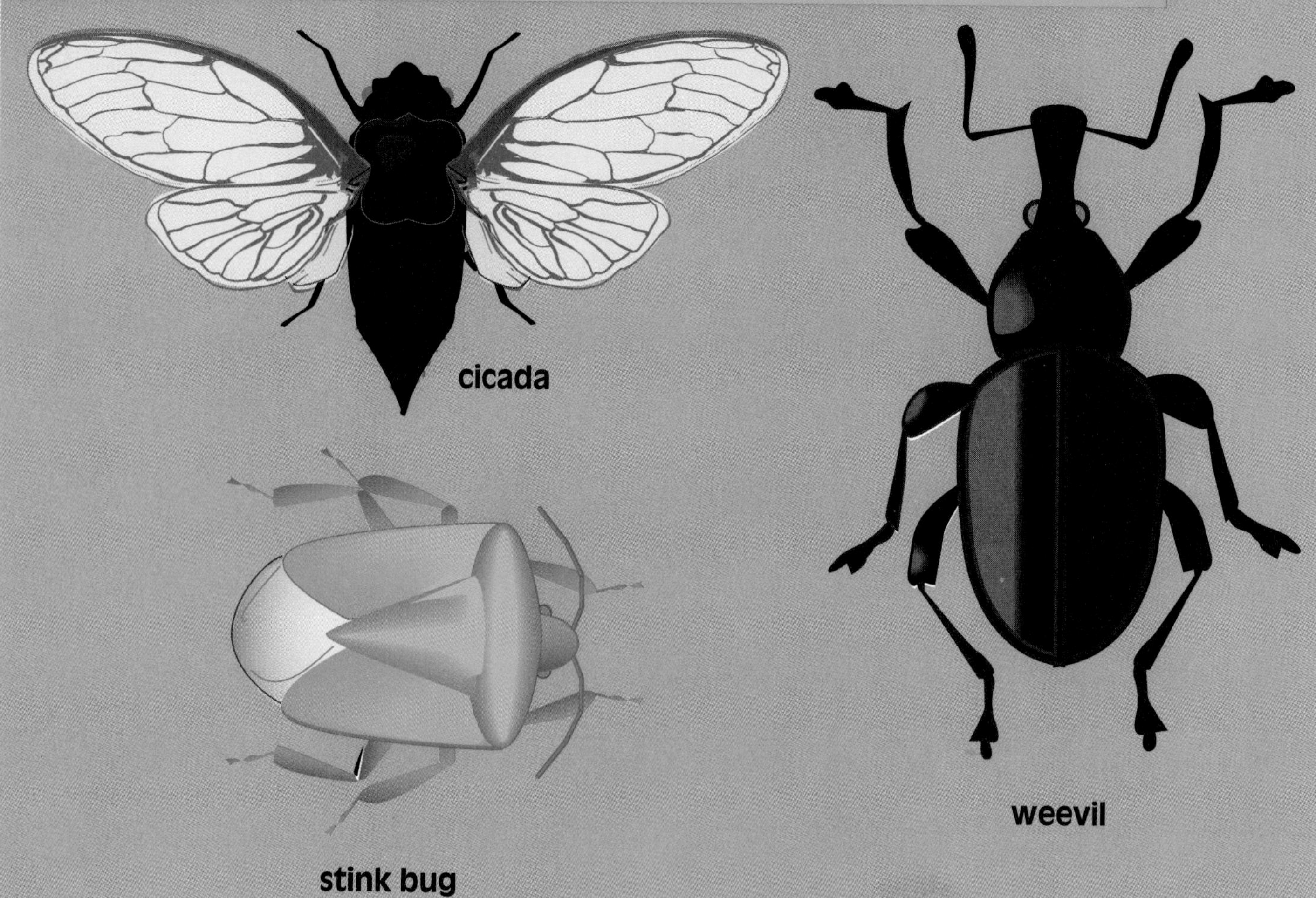

The Vertebrates

Vertebrates are animals that have backbones. The vertebrates include more than 45,000 different ***species*** divided into five groups—fish, amphibians, reptiles, birds, and mammals. Vertebrates have a protective skin covering, an inside skeleton, muscles, blood that circulates through blood vessels, lungs or gills, and most have fins or legs. They also have heads with brains that are part of a well-developed nervous system. The nervous system processes information from sense organs, such as the eyes and the nose.

The Cold-Blooded Truth

Fish, amphibians, and reptiles are called cold-blooded, while birds and mammals are called warm-blooded.

Cold-blooded means that the body temperature of an animal depends on the temperature of its surroundings. If a cold-blooded animal is too cold, it cannot move very fast and will search for a warmer place. If it is too warm, it will move to a cooler place or move its body to cool off.

Warm-blooded means that the body temperature of an animal remains much the same, no matter if its surroundings are hot or cold. Birds and mammals burn fat reserves to heat their bodies, and most warm-blooded animals sweat to cool off. Although you might want to wear a coat when it's cold outside, your inside body temperature remains at about 98.6°F.

snakes are cold-blooded

cats are warm-blooded

Fish

Covered with scales or bony plates, live in salt or fresh water, breathe through gills, and use fins to swim. They are ***cold- blooded*** (see p. 11), and their young hatch from eggs. Sharks, rays, and dog-fish have skeletons made of ***cartilage***, a rubbery substance. Other fish have skeletons of bone. Fish come in an amazing variety of sizes and shapes, from sea horses to catfish, balloonfish to swordfish, and flying fish to eels. In all, scientists have identified more than 20,000 species.

Amphibians

Most lay eggs in water. Young tadpoles breathe through gills and are like fish in other ways, too. Adults develop lungs like those of land animals and often live on land. They take in additional oxy-gen through their thin, smooth, moist, scaleless skin. Amphibians are ***cold-blooded*** (see p. 11). Most are ***insectivores***, meat eaters whose diet consists mainly of insects. Frogs, toads, salamanders, and newts are among the 2,400 different amphibian species.

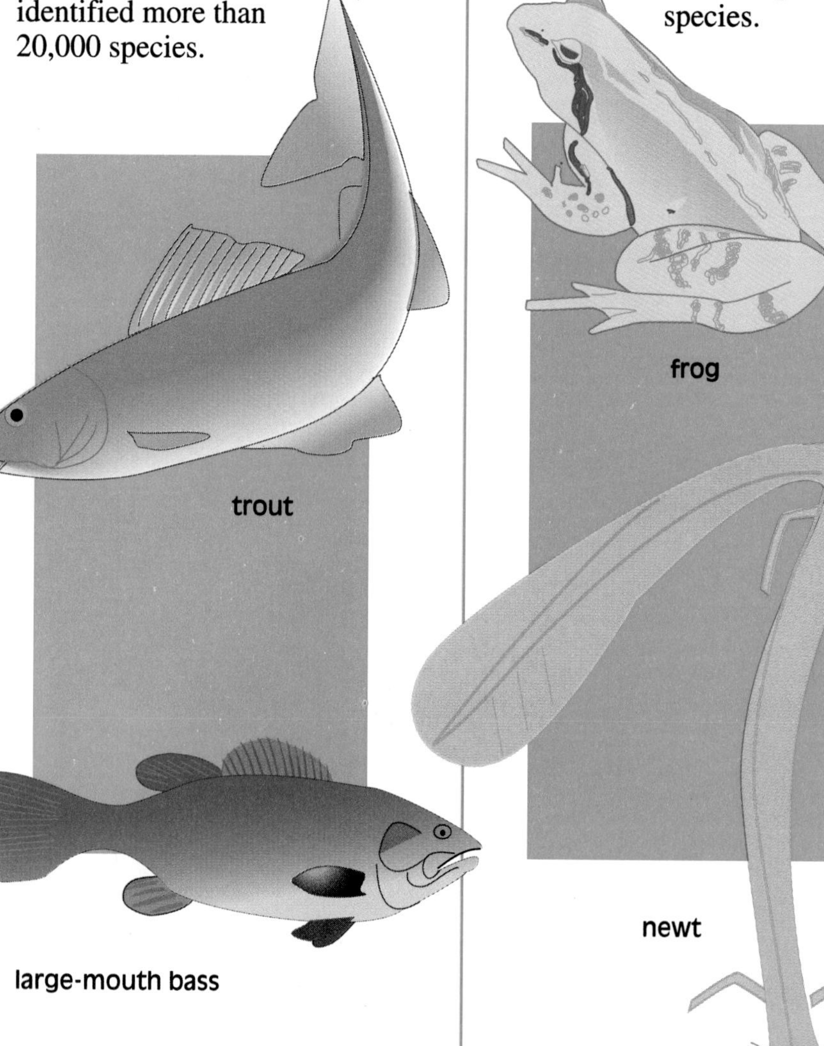

Reptiles

The first true land animals, rep-tiles do not have to return to water to reproduce. Their eggs have tough, leathery shells. Reptiles are covered with dry scales, breathe through lungs, and most are ***cold-blooded*** (see p. 11). They are both ***herbivores*** and ***carnivores*** (see p. 18). Most have four legs with five claws on each foot. The 6,000 known reptile species are divided into four groups:
1. Crocodiles, alligators, gavials and caymans—large animals with powerful tails; 2. Turtles and tor-toises—animals with hard bony shells; 3. Lizards and snakes; and 4. Tuatara, a creature with three sets of eyelids found only in New Zealand.

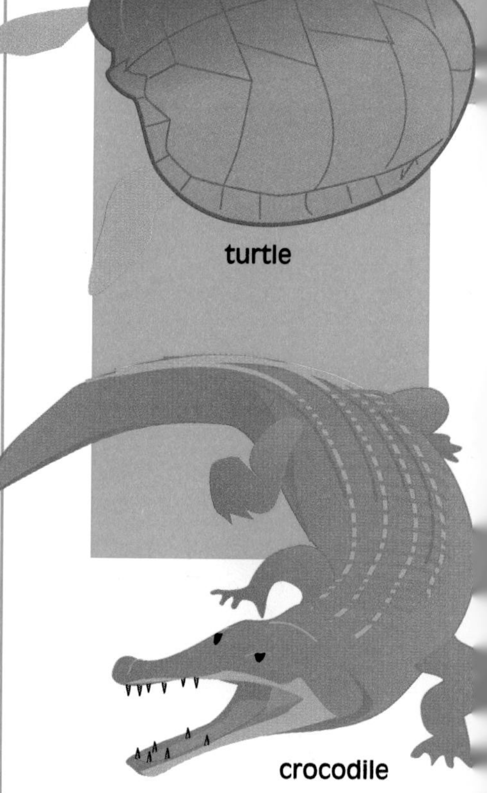

Birds

The only animals with feathers, wings instead of front legs, and two feet for walking. Birds have a bill and no teeth. They have excellent eyesight, but a bad sense of smell. Most birds build nests where eggs are laid, protected, incubated, and hatched. The nest provides a home for baby birds until they ***fledge***, or learn to fly and live on their own. Birds are ***warm-blooded*** (see p.11). Birds' bills, plumage, body shape, and size vary to suit their habitats—from tropical rain forest to freezing tundra—and eating habits—***carnivore*** and ***herbivore*** (see p. 18). Among the more than 9,000 known species are tiny hummingbirds, flightless ostriches, parrots, and penguins.

Mammals

The only animals with hair. Mammals don't lay eggs but bear their young live. Female mammals have mammary glands to produce milk and nurse their young. All mammals have three different kinds of teeth (incisors, molars, and canines), and are ***warm-blooded*** (see p.11). Mammals are ***carnivores***, ***herbivores***, and ***omnivores*** (see p. 18). Among the 4,500 known mammal species are mice, elephants, whales, chimps, giraffes, hamsters, and humans.

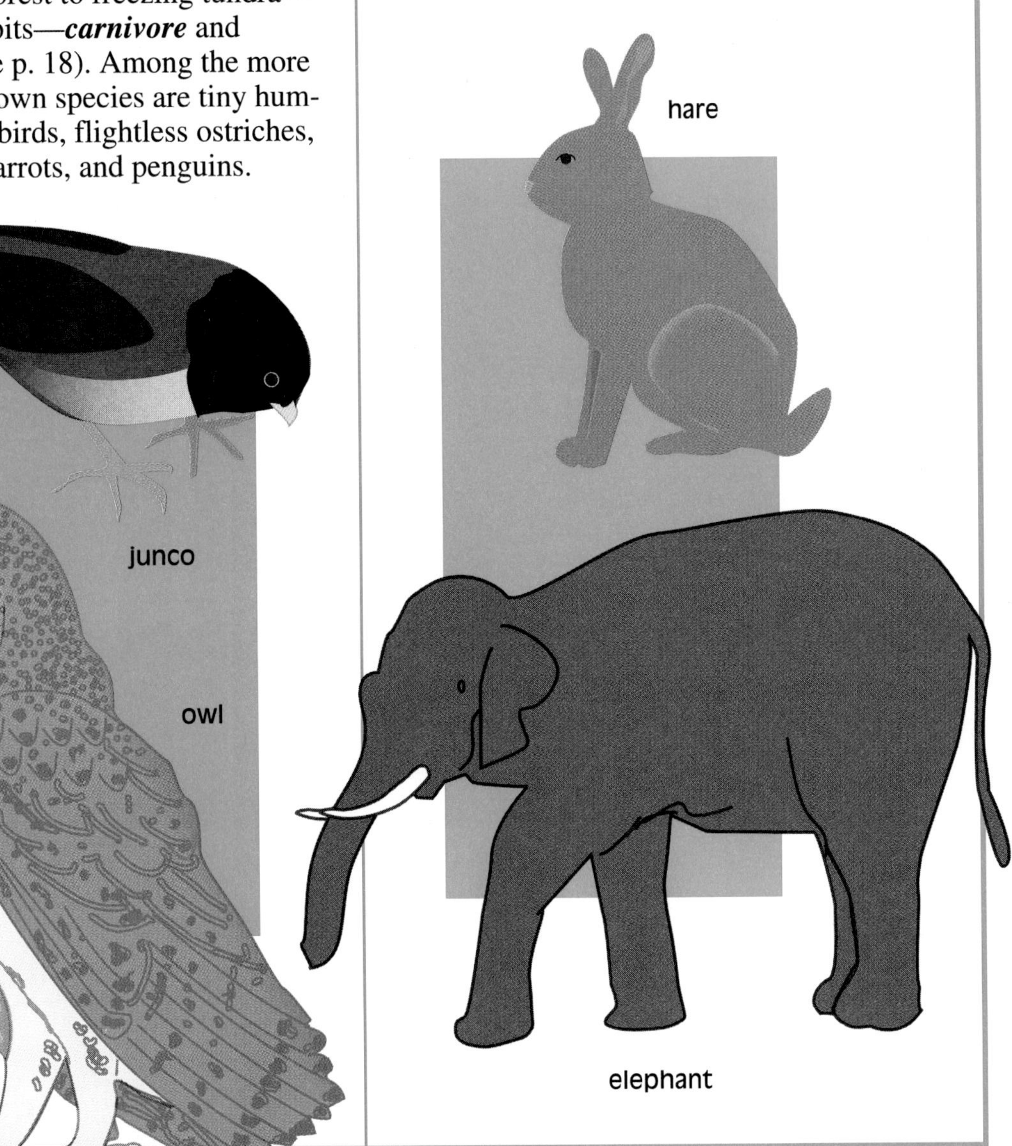

TALKING IT OVER:
Animal Communication

Unlike humans, other animals don't sit down and talk things over. Instead, they communicate using signals and calls. Each species of animal has its own special way of communicating.

Communicating is an important part of ***animal behavior*** (see pp. 38-39). Calling, croaking, chirping, screeching, flapping wings, ducking heads, raising tails, and releasing chemicals are just some of the ways that animals warn each other about danger, attract mates, alert neighbors to a source of food, mark their territories, and, in some species, even announce death.

The male peacock displays his beautiful tail feathers to attract peahens.

Animal Relatives

How can you tell animal cousins apart? Here are some clues:

Crocodiles and Alligators

Crocodiles have long, pointed snouts that are wider at the eyes and narrower at the nose. Alligators have rounded snouts, as wide at the eyes as at the nose.

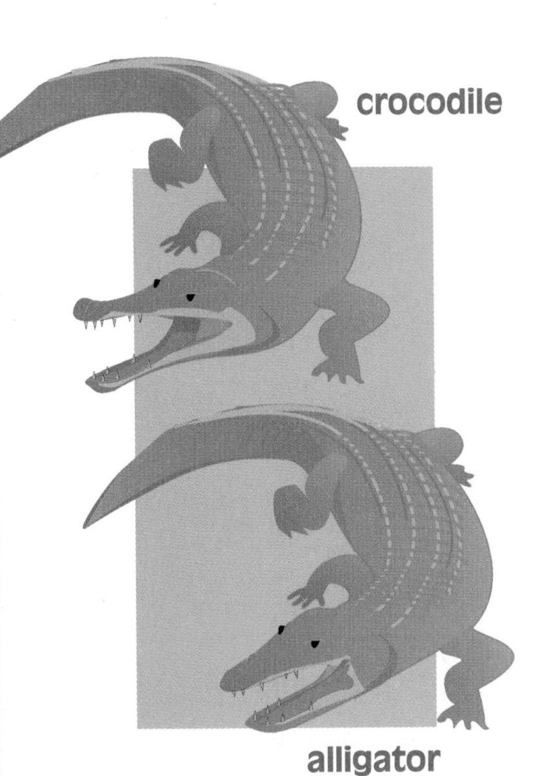

Frogs and Toads

Frogs have smooth skin. Toads have warty skin.

Turtles and Tortoises

Turtles live in fresh water and salt water, and on land. Tortoises are turtles that live only on land.

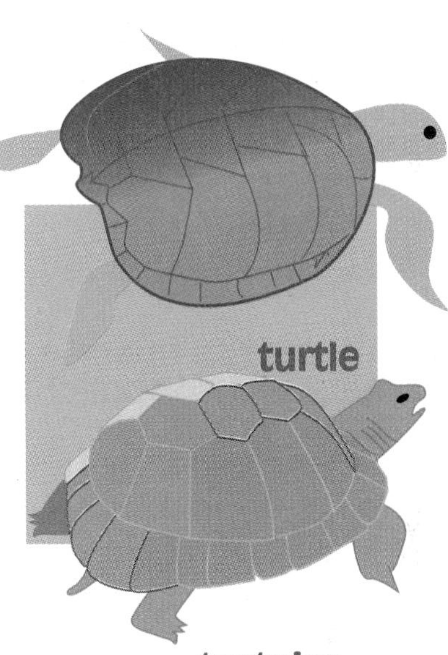

African Elephants and Asian Elephants

African elephants are bigger than Asian elephants. They have bigger ears and a flatter back. African elephants cannot be tamed, but people have ridden around on Asian elephants for centuries.

Dinosaurs

Dinosaurs appeared on earth about 230 million years ago. They roamed the land for nearly 160 million years before the last great dinosaurs died out, about 65 million years ago. Why dinosaurs became extinct is a mystery, although most scientists agree that a change in the earth's climate, animals, and plants probably led to the dinosaurs' death.

Some Possible Reasons Why Dinosaurs Disappeared

1. **A natural disaster, such as an earthquake or a volcanic eruption, destroyed the dinosaurs and their habitat.**

2. **A plague of smaller animals preyed on dinosaur eggs, eating so many that the dinosaur population became too small to survive.**

3. **A long dry period, drought, or other change in plant life left dinosaurs with inadequate food sources.**

4. **A giant meteorite ripped into the earth, raising a dust cloud that blocked sunlight and brought about a sudden change in climate—one too cool for the dinosaurs and the many other plants and animals that died out at about the same time.**

Dinosaurs were reptiles. Some walked on two legs, others on four. We know about dinosaurs from their ***fossil remains*** (see p. 47). Fossils allow us to trace the development of dinosaurs throughout the Mesozoic Era, the geologic time period that the dinosaurs dominated.

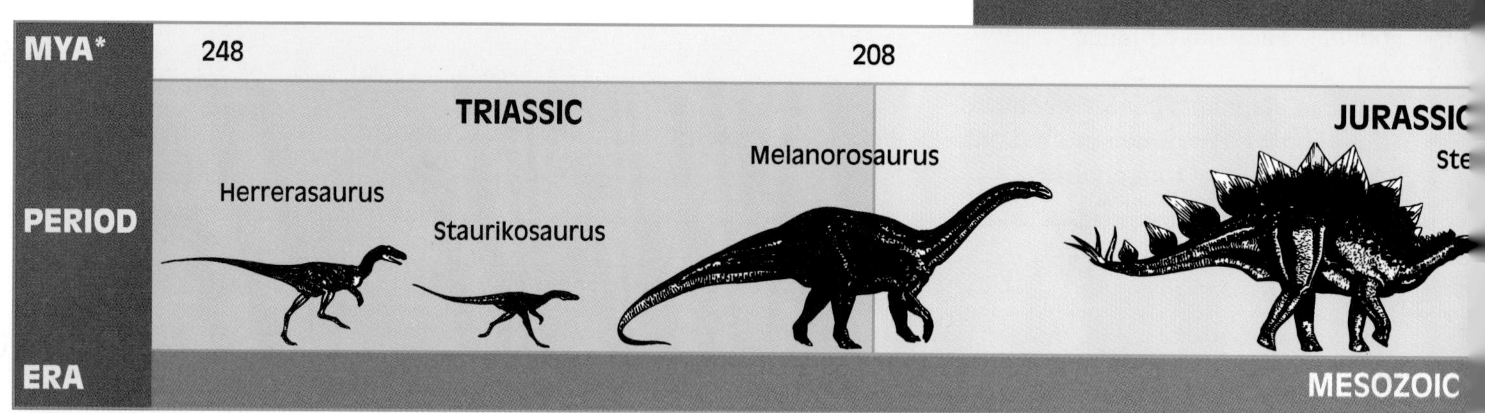

* MYA means *million years ago*

144

65

CRETACEOUS

Tyrannosaurus

…imus

Velociraptor

Torosaurus

Struthiomimus

ANIMAL CHAMPIONS

Fastest Land Animal	Cheetah	70 miles per hour
Fastest Diver Through Air	Peregrine Falcon	115 miles per hour
Fastest Flyer	Common Eider	45 miles per hour
Deepest Diver Through Water	Emperor Penguin	870 feet
Slowest Land Animal	Snail	2-3 feet per minute
Largest Animal	Blue Whale	150 tons
Largest Land Animal	African Elephant	5 tons
Longest Snake	South American Boa Constrictor	37 feet long

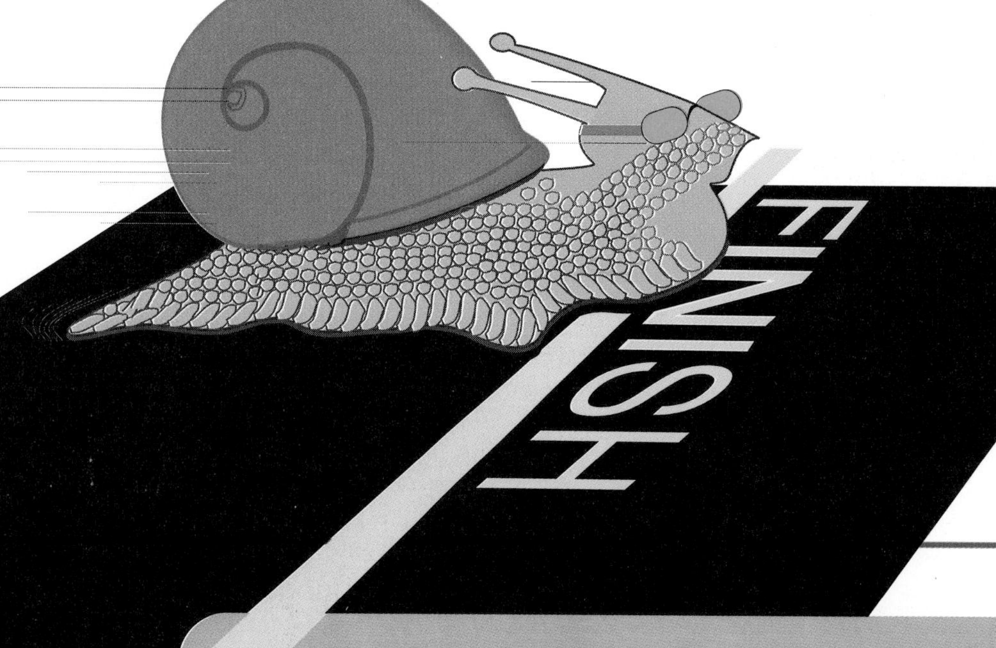

Who Eats What or Whom?

Carnivores eat animal flesh (meat). Lions, sharks, dogs, cats, dolphins, and seals are all carnivores.

Herbivores eat plants. Many tiny animals eat only plants to survive. So, too, do elephants, giraffes, cows, horses, blue whales, and other huge animals.

Omnivores eat both meat and plants. Bears, humans, and raccoons are omnivores.

Predators are carnivores that hunt other animals, or ***prey***, for food. Lions, eagles, alligators, many snakes, cheetahs, owls, and many other animals hunt live prey.

Scavengers eat dead animals or the discarded remains of predators. Hyenas and vultures are well-known scavengers.

How Animals Reproduce

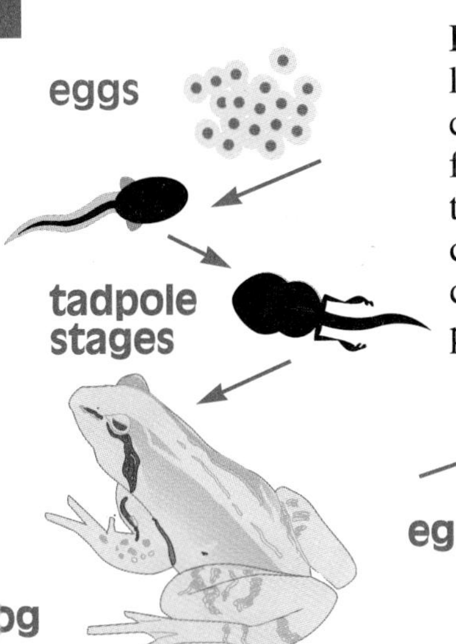

Butterflies lay eggs on leaves. These grow into caterpillars. Caterpillars form a chrysalis and enter the pupa stage. In the chrysalis, ***metamorphosis***, or change, takes place. The pupa emerges as a butterfly.

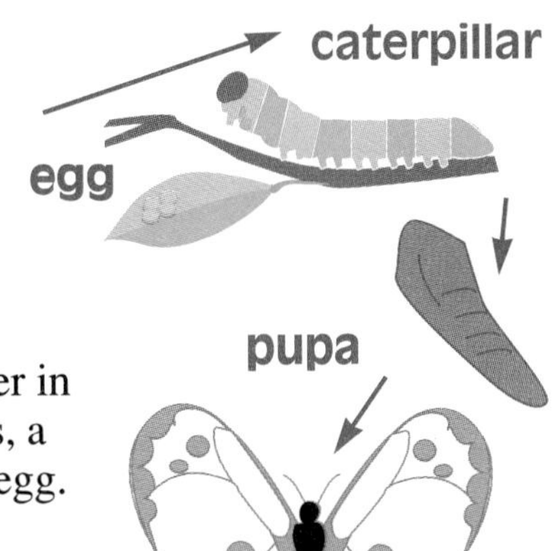

Frogs lay eggs underwater in springtime. In two weeks, a tadpole hatches from an egg. At nine weeks, front and back legs form, and by the eleventh week, an adult frog is formed.

Mammals always give live birth. In most mammals, eggs develop inside the mother's body.

Juvenile birds are hatched from eggs. Like many other animals, young birds grow bigger, but hatch in their basic adult form.

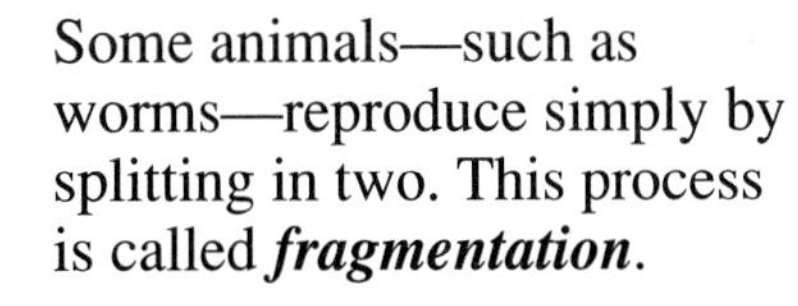

fragmentation

Some animals—such as worms—reproduce simply by splitting in two. This process is called ***fragmentation***.

BABY ANIMALS

Animal	Baby's Name	Animal	Baby's Name
Bear	Cub	Goose	Gosling
Beaver	Kit	Hare	Leveret
Bird	Nestling	Hawk	Eyas
Cow	Calf or Heifer	Horse	Colt
Deer	Fawn	Kangaroo	Joey
Duck	Duckling	Lion	Cub
Eagle	Eaglet	Rooster	Cockerel
Eel	Elver	Seal	Pup or Whelp
Elephant	Calf	Shark	Cub
Fish	Fingerling or Fry	Swan	Cygnet
Fox	Cub or Kit	Turkey	Poult
Giraffe	Calf	Whale	Calf
Goat	Kid	Zebra	Colt

THE PLANT KINGDOM

1 Plants

More than 400,000 different plant species have been identified growing both on land and in fresh and salt water. Most life on the earth would not exist without plants. They are a food for many animals and they give off the oxygen that animals need to breathe. Besides food and oxygen, plants provide people with medicines, building materials, textiles (fabric for clothing), and other useful things.

The First Plants

Plant life first developed in the seas and oceans that covered the earth in ancient times. Then, about 425 million years ago, plants began to invade the land. Among early land plants that are still alive today are mosses, liverworts, and horsetails.

Plants are divided into two groups—those that make seeds and those that do not.

Seed Plants	
	Angiosperms (Flowering Plants) Monocot
	Dicot
	Gymnosperms (Naked Seed Plants) Conifers and others

Some Plants That Don't Make Seeds	
	Moss
	Liverwort
	Horsetail
	Fern

Description	Examples	
Plants produce seeds with one food part (cotyledon), leaves have parallel veins	Grasses, palms, lilies, and orchids	wheat
Plants produce seeds with two food parts (cotyledons), leaves have branching veins	Oak trees, roses, broccoli, tomatoes, and many others	bean plant
Plants produce exposed seeds, usually on cones, and have no flowers. Most have needlelike leaves and are evergreen	Pines, spruces, junipers, firs, and other conifers, yews, and ginkos	conifer branch with pine cone
Produces spores, tiny green plants, no true roots, stems, or leaves, needs moist place to live		moss
Produces spores, tiny green plants, no true roots, stems, or leaves, needs moist place to live		liverwort
Produces spores, no true roots, stems, or leaves, needs moist place to live		horsetail
Produces spores, green leafed, needs moist place to live		fern

Making Seeds: Conifers and Flowering Plants

Seed plants are divided into two groups—***gymnosperms***, or cone-bearing plants, and ***flowering plants***, including monocots and dicots.

Storing Food: Monocots and Dicots

In flowering plants, food is stored in one or two "seed leaves" called ***cotyledons.*** Seed leaves with one food part are called ***monocotyledons***, or ***monocots***. Seed leaves with two food parts are called ***dicotyledons***, or ***dicots***.

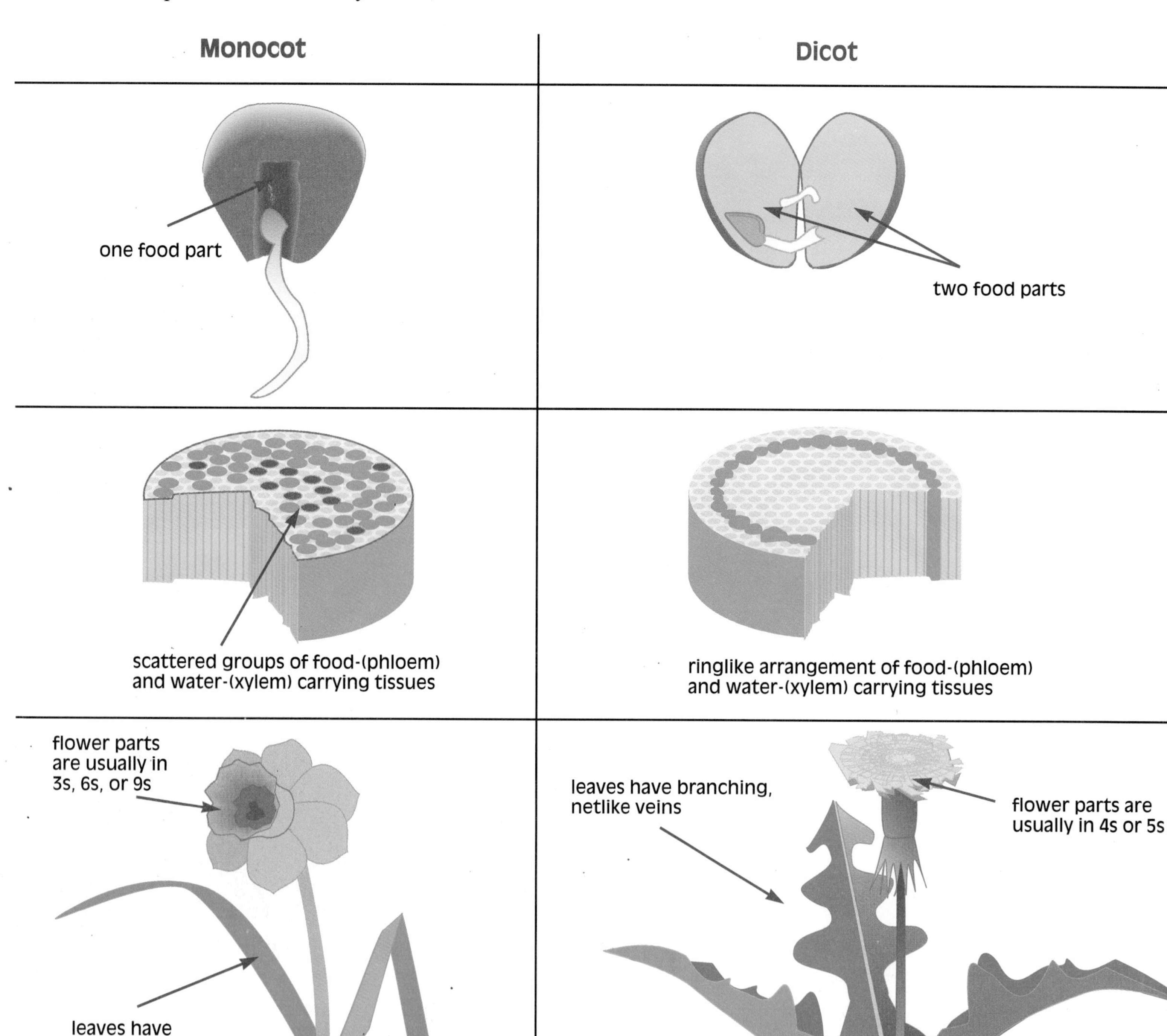

Plant Parts

Leaves

Leaves are the plant's food factory.
In ***photosynthesis*** (see p. 28), ***chlorophyll***, a green substance, captures energy from the sun. The energy is used to combine water from the soil and carbon dioxide gas from the air to make sugar. Oxygen is released during the process. During ***respiration*** (see p. 29), the plant burns sugar to nourish itself. Carbon dioxide is released during the process.

Plants sweat.
Water coming up from the roots is released into the air in the form of water vapor in a process called ***transpiration*** (see p. 29). Plants sweat only on the undersides of their leaves.

Stems

Stems are the plant's support system.
Stems hold leaves up to the light and keep fruits and flowers attached. They grow straight up, trail along the ground, climb fences and trees, or stay underground.

Stems are also a food and water highway.
Water and minerals from the soil are transferred upward from the roots through the ***xylem*** tissues in the stem to leaves and other plant parts. Food made in the leaves is transported through the ***phloem*** tissues in the stem to the growing parts of the plant. The pumping of water and minerals is called ***capillary action.***

Some stems are food storage sites.
White potatoes are really a type of underground storage stem called a ***tuber***.

Roots

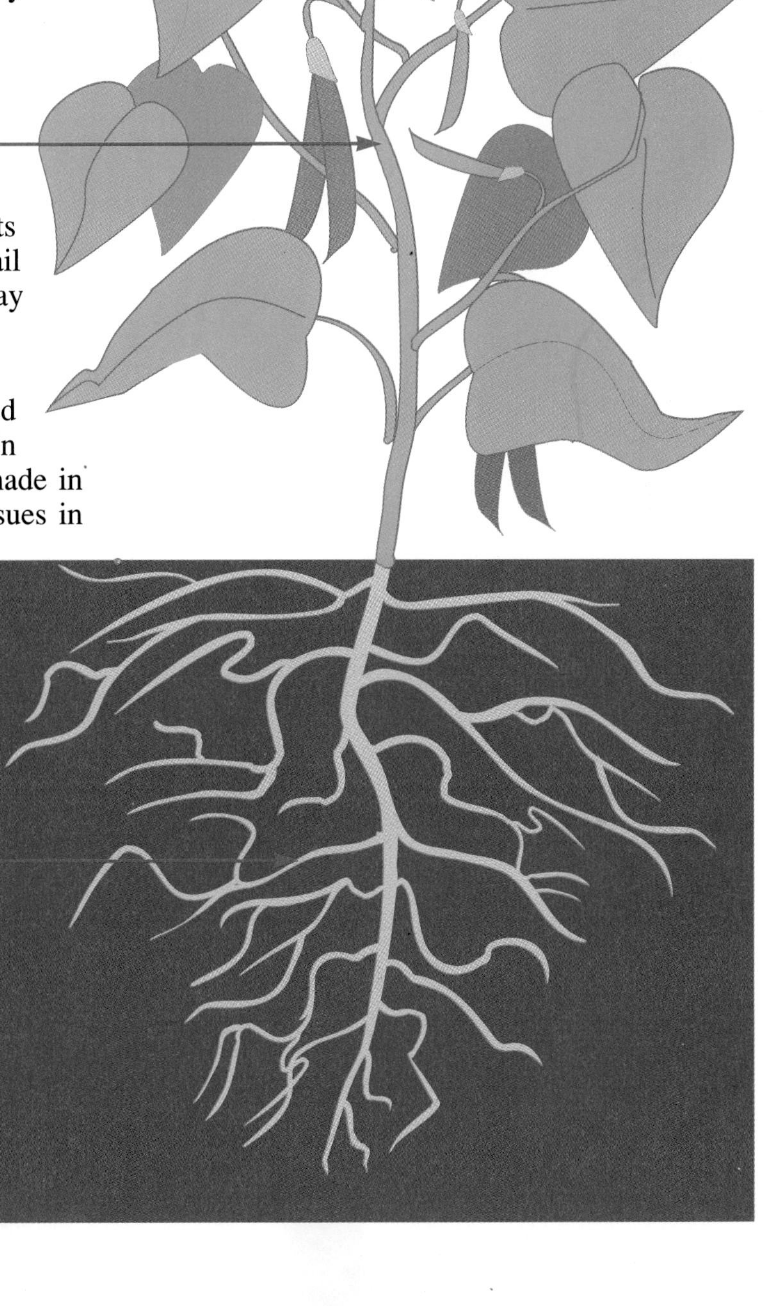

Roots are the plant's anchors.
Roots hold the plant firmly in the ground.

Roots are also absorbers.
Tiny hairs take in water and minerals from the soil.

Roots provide storage.
Plants transport sugars and starches to the root for storage. Carrots, beets, and radishes are some of the root foods we eat.

Woody and Nonwoody

Trees, shrubs, and many vines have woody stems. The older part of the woody stem, or ***trunk***, grows outward each year, increasing the plant's thickness. The tip grows upward, making the plant taller.

Herbaceous plants have nonwoody stems. They usually have green, supple stems and do not have bark. Like woody plants, herbaceous plants have xylem and phloem tissues to transport water and food. These are arranged differently in ***monocots*** and ***dicots*** (see p. 22).

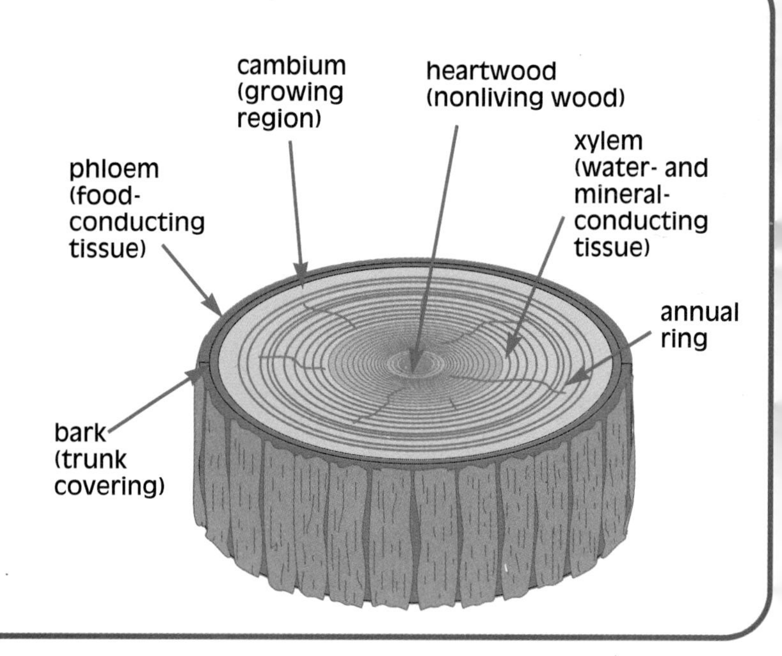

Leaves and Leaf Parts

How Plants Reproduce

Each plant and animal has a ***life cycle***, a pattern of growth, reproduction, and death.

Reproduction by Seeds

Most plants reproduce through seeds. ***Seeds*** are the fertilized ovules, or plant "eggs," from which new plants grow. Some seeds ***germinate***, or start growing, as soon as they leave the parent plant. Others can take weeks or even months to germinate. Some seeds are eaten by birds and animals. There seeds pass through the animal's digestive tract and are eliminated as waste, often to sprout far from their parent plant.

Both conifers and flowering plants produce seeds.

Conifers

Conifers make seeds in cones. Most conifers have two kinds of cones, a male cone that contains ***pollen*** and a female cone that contain ***ovules***. Pollen contains cells that ***fertilize*** ovules so that they become ***seeds*** for new plants.

To make seeds, ***pollination*** and ***fertilization*** must occur. In most conifers, wind carries pollen from pollen cones to ovule cones. Fertilization occurs when the pollen lands on an ovule, fertilizing it. The fertilized ovules are seeds that contain a new plant embryo.

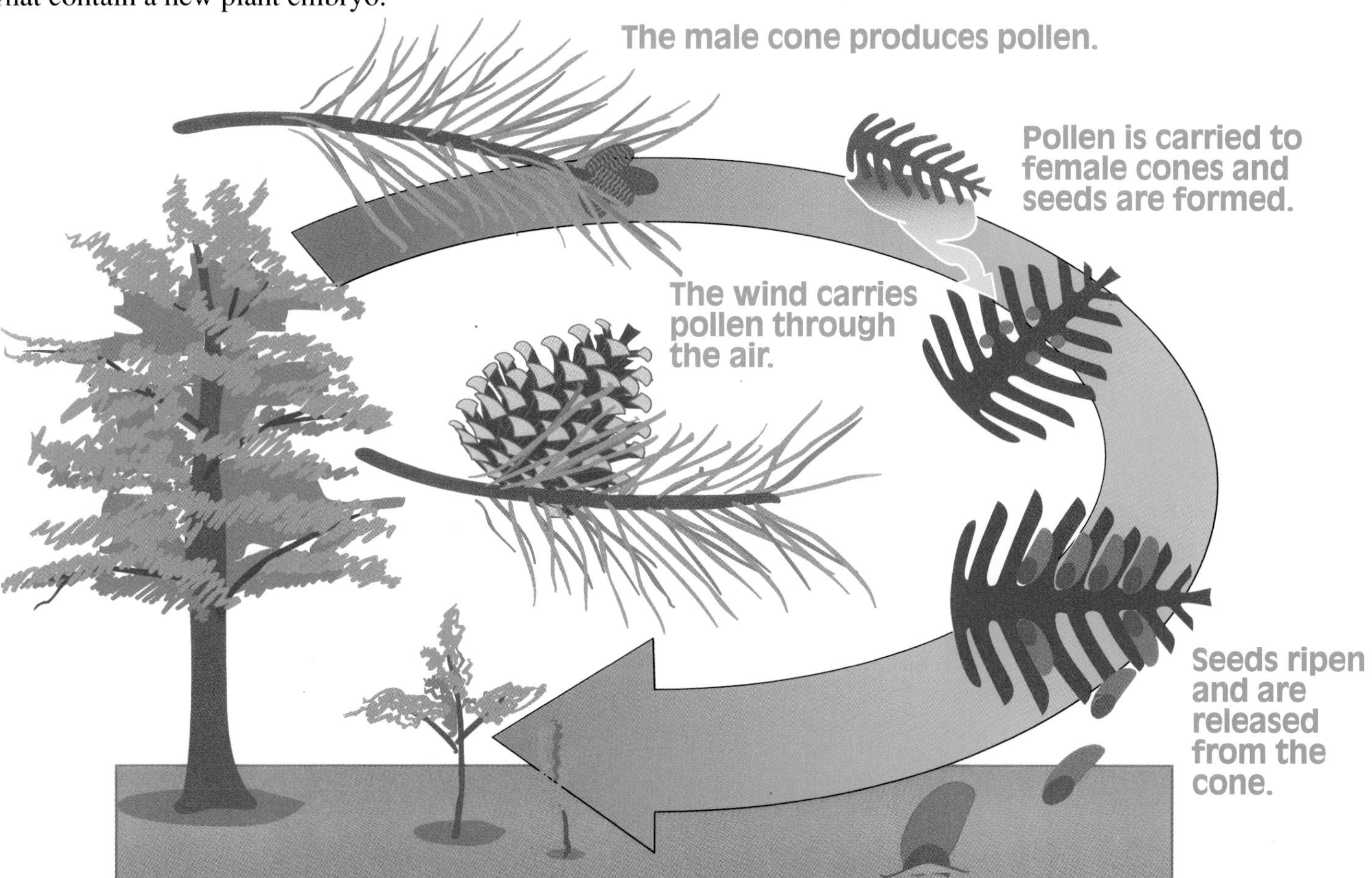

The male cone produces pollen.
Pollen is carried to female cones and seeds are formed.
The wind carries pollen through the air.
Seeds ripen and are released from the cone.
Seeds germinate and become new conifer plants.

Flowering Plants

Flowering plants make seeds in ***flowers***. Most flowers contain both a male part called a ***stamen*** and a female part called a ***pistil.*** These are called ***perfect*** flowers. Plants with separate stamen flowers and pistil flowers are said to have ***imperfect*** flowers.

Pollination

Pollen is produced in the "male" stamen of a flower. The plant's ***ovules***, or "eggs," are produced in the ovary of the "female" pistil of a flower. To fertilize the ovules, the pollen travels from the stamen's ***anthers*** to the pistil's ***stigma***, the flower part that catches pollen.

Pollen can be blown by the wind or carried by animal pollinators, including hummingbirds, bees, butterflies, beetles, and other insects. Flowers often have special colors, shapes, or markings to attract pollinators.

Fertilization

Fertilization occurs when pollen enters an ovule in the ovary. After a pollen grain lands on the stigma, it sprouts a tube. The tube grows down the ***style***, the slender neck that connects the stigma to the ovary. The pollen tube then enters the ovule, allowing the male plant cells to ***fertilize*** the female plant cells. The fertilized ovule develops into a ***seed*** containing a plant embryo.

Seed Protection

Seeds are protected by ***fruit***. Fruit is formed from the ovary surrounding the seed. Moist, fleshy fruits include peaches, grapes, tomatoes, and melons. Dry, hard fruits include coconuts, walnuts, and pea pods.

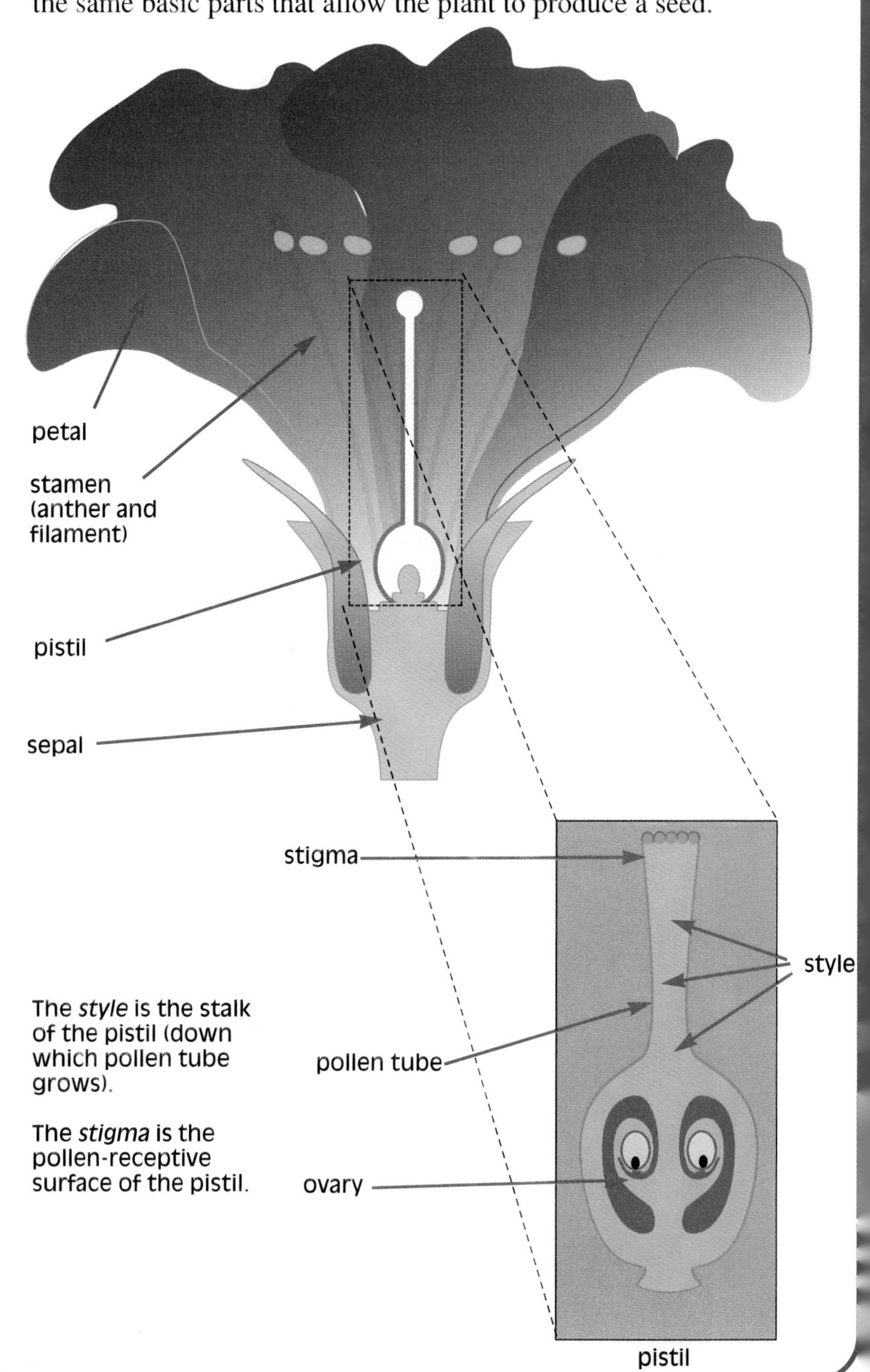
Parts of a Flower
Flowers come in many shapes, sizes, textures, and colors. But all flowers are reproductive structures, and all flowers are made up of the same basic parts that allow the plant to produce a seed.
petal
stamen (anther and filament)
pistil
sepal
stigma
style
The *style* is the stalk of the pistil (down which pollen tube grows).
The *stigma* is the pollen-receptive surface of the pistil.
pollen tube
ovary
pistil

WHO NEEDS SEEDS? Reproduction from Plant Parts

Many plants can grow from plant parts, not just from seeds. Plants that have been hurt or cut can sprout new growth from damaged stems, roots, or even leaves. The process of growing new plants from existing plant parts is called ***vegetative reproduction***. Plants use a variety of parts to produce new plants:

- **Tubers, bulbs, and corms.** All are types of underground stems. The "eyes" or buds of tubers, such as white potatoes, grow into roots and shoots to produce a new plant. Bulbs, such as onions, are big buds made up of a stem and special types of leaves—the onion layers. Corms, used to grow such flowers as gladiolus, are fleshy stems.
- **Runners and rhizomes.** These stems run along the ground (runners) or spread underground (rhizomes). New strawberry plants grow from the tips of runners. Many grasses and weeds are spread by rhizomes.
- **Roots**. Some fruit trees and bushes, such as apples and raspberries, send up "suckers," or new shoots from the roots. Roots of some plants, such as dandelions, can produce new plants from small, broken-off root pieces.
- **Leaves**. Some plants, such as the houseplant kalanchoe, produce little plantlets right on their leaves. African violets and some other plants produce plants from leaves placed on top of soil.

Reproduction Without Seeds

Spores

Ferns, mosses, horsetails, and other seedless plants reproduce by ***spores***. Spores are reproductive cells—tiny bits of plant life. Ferns, for example, produce spores on a parent plant, where they are stored in spore cases until they are ripe. Then the parent plant releases them. In most living ferns, the spores grow into a tiny, heart-shaped ***prothallus***, which produces sperm cells and eggs. Water must be present so that the sperm can swim to an egg and fertilize it. The embryo grows into a ***sprororyte***, or new fern plant.

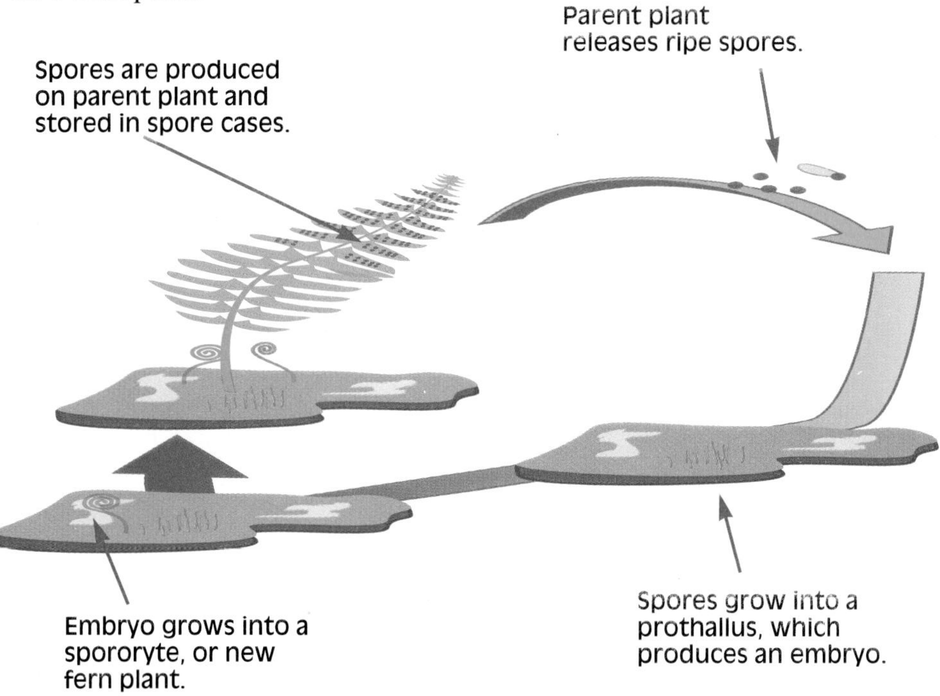

Seedy Stuff:
The Inside of a Seed

Inside every seed is:

1. **A tiny new plant. The new plant is called an embryo.**
2. **Food to nourish the plant. Each seed has either one or two cotyledons or food storage areas (monocots have one, dicots have two).**

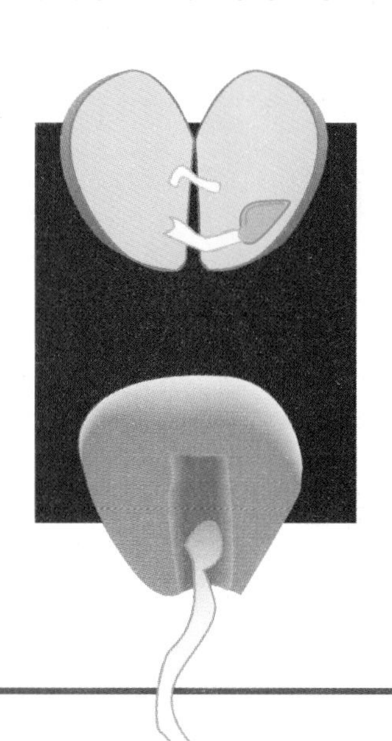

4 Photosynthesis, Respiration, and Transpiration

Photosynthesis

In ***photosynthesis***, plants use the sun's energy to make sugar. Special leaf structures called ***chloroplasts*** contain a green pigment called ***chlorophyll.*** The pigment absorbs light energy from the sun. This energy is used to combine water from the soil and carbon dioxide gas from the air into sugar. During this reaction, oxygen is given off and released into the air.

Photosynthesis is responsible for the oxygen we breathe.

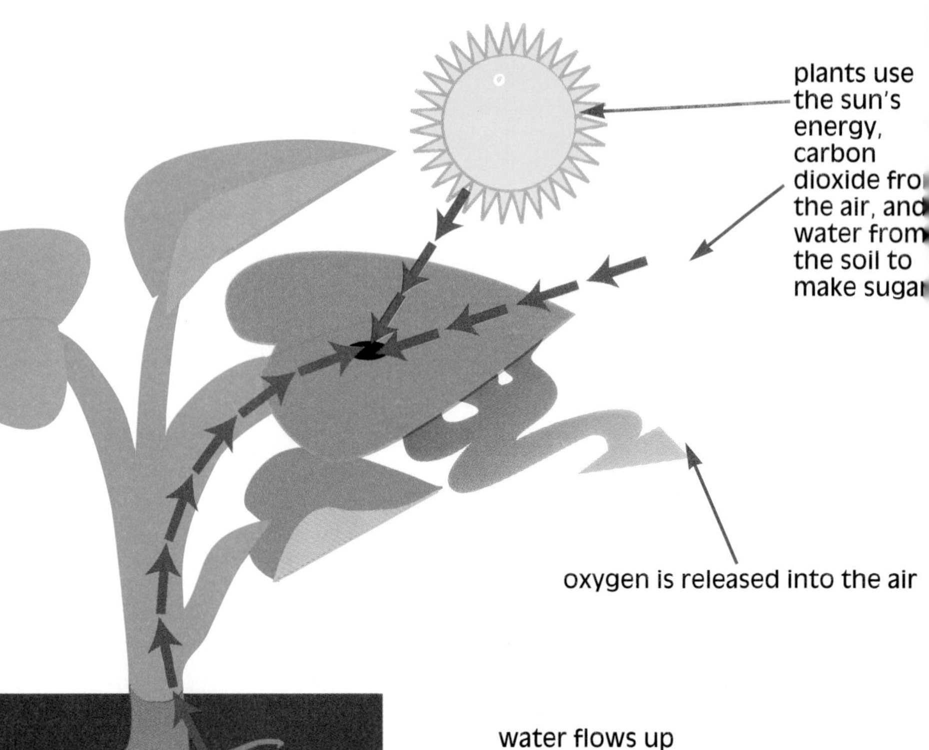
plants use
the sun's
energy,
carbon
dioxide fro
the air, and
water from
the soil to
make suga
oxygen is released into the air
water flows up
through the roots
from the soil

That's a Lot of Water!

As much as 168 gallons of water ***transpire*** from one corn plant during a single growing season. If you lost that much water, you'd have to drink almost 2,000 gallons over the same period to remain healthy!

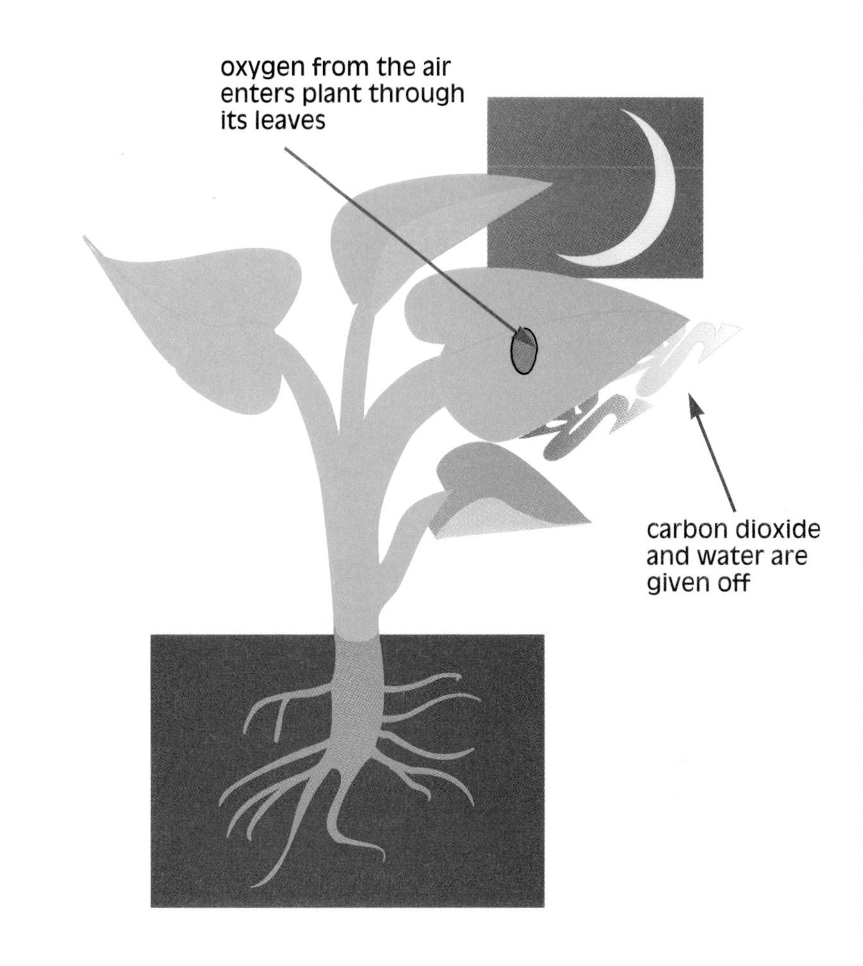
oxygen from the air
enters plant through
its leaves
carbon dioxide
and water are
given off

Respiration

Plants use the food they make during ***photosynthesis*** to grow and maintain themselves. Food is stored energy. The process of breaking down food so that its energy can be used is called ***respiration.*** In this process, cells use oxygen from the air to release the food energy. Carbon dioxide and water are given off during the reactions. In many ways, respiration is the opposite of photosynthesis.

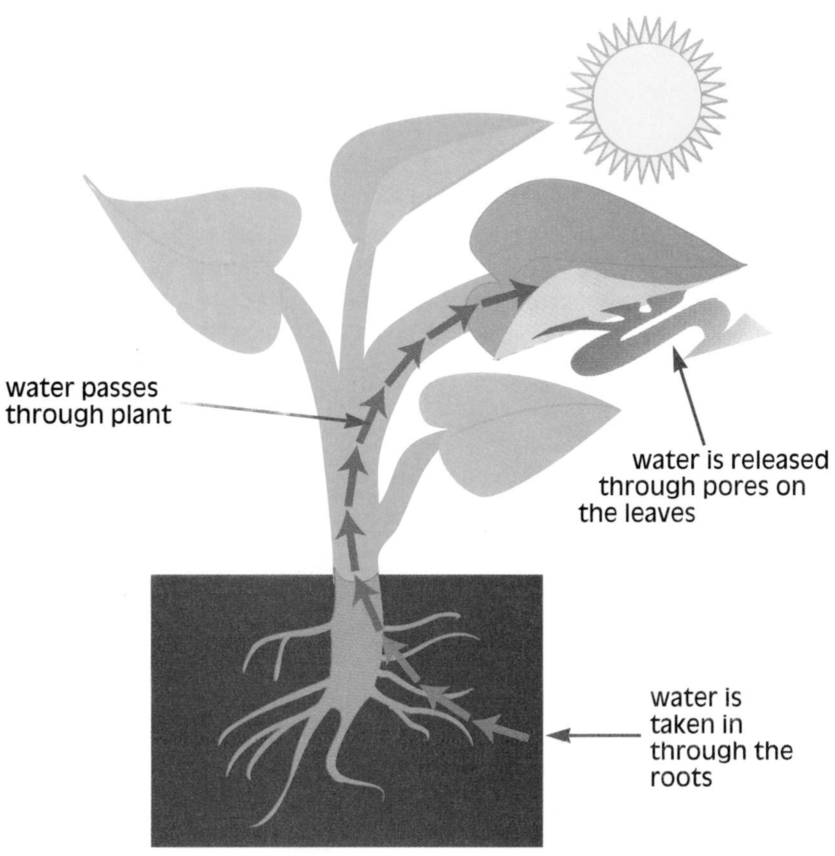
water passes
through plant
water is released
through pores on
the leaves
water is
taken in
through the
roots

Transpiration

Most of the water a plant takes in through its roots passes through the plant and out the pores on the undersides of the leaves. The water then evaporates into the air. This process is called ***transpiration.***

ECOLOGY AND THE LANDSCAPES OF LIFE

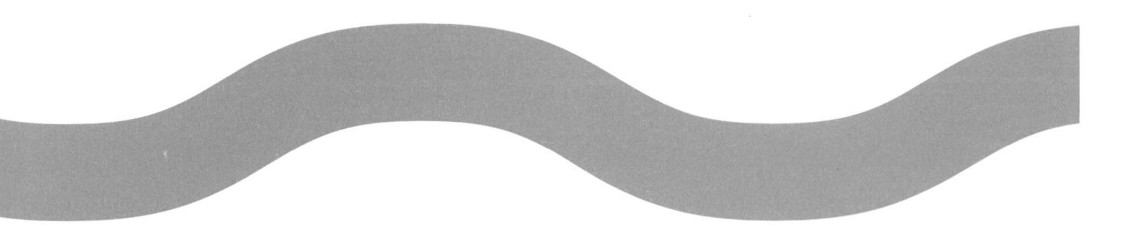

1 Biomes

Biomes are large areas or ***environments*** that share the same general climate of temperature and rainfall. Different biomes support different types of plants and animals.

Grasslands

Grasslands are farming and grazing biomes. They are known by many names throughout the world—***prairies, savannas, steppes,*** and ***pampas.*** Rainfall averages ten to forty inches a year, enough to grow grasses, but not enough to produce thick forests. The soil is deep and rich. Grasslands usually occur in the interior parts of continents. Fire, periodically set by lightning or humans, releases nitrogen and other minerals into the soil and helps certain seeds to germinate.

Traditionally, great herds of grazing animals—bison, wildebeest, antelopes, zebras, and others—have moved across the grassy plains and savannas, followed by predators and scavengers—lions, wolves, vultures, hyenas, and many more. Now, many of the world's grasslands, including those in the United States, are farmlands where wheat, barley, oats, rye, and other grains grow. Cattle graze where wild animals used to roam.

Deserts

Desert biomes usually receive less than ten inches of rain per year. In most deserts, temperatures vary greatly, from extremely hot days to cool, even freezing nights.

Plant and animal life in deserts is often scarce and is well adapted to the harsh environment. Cacti and euphorbia are typical desert plants. Called ***succulents***, these plants store water in their waxy leaves and stems, which slow ***transpiration***, or water loss (see p. 29).

Animals in desert biomes are often ***nocturnal***, active only at night. These animals burrow deep into the earth during the day to escape the heat and sun. Certain types of lizards and snakes thrive in the hot desert, as do animals that need only a little water.

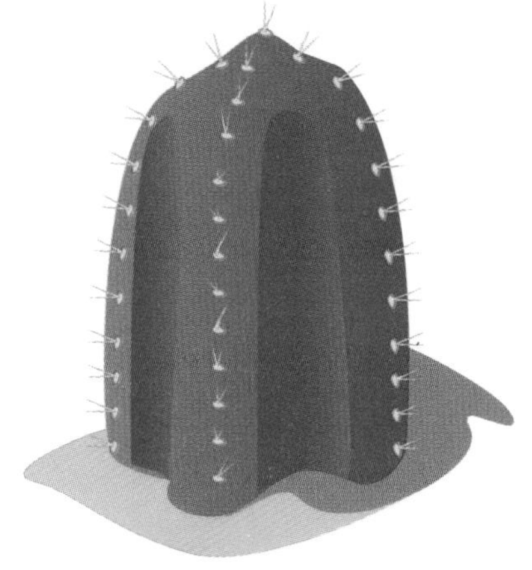

Scrublands

Scrubland, or ***chaparral***, is a coastal biome that is cool and moist in winter and hot and dry in summer. Most scrublands average around ten inches of rain per year, and almost all of it falls during winter months.

Drought-resistant evergreen shrubs provide cover and homes for lizards, rodents, rabbits, and their predators. During winter months, larger animals, such as deer, move in to nibble on the leaves.

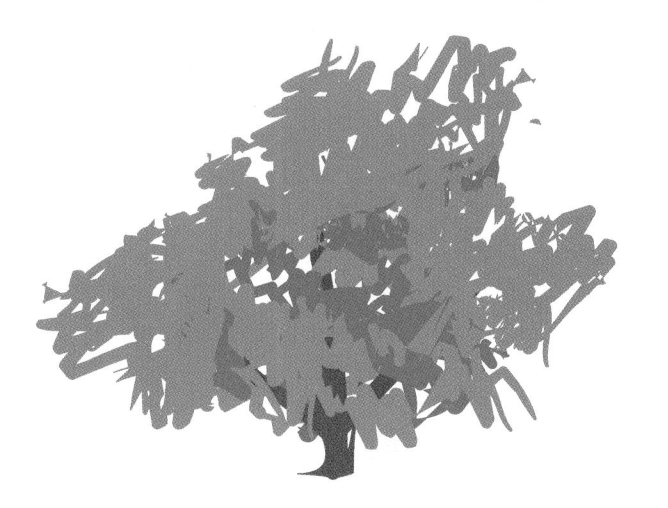

Deciduous Forests

Deciduous forests are dominated by deciduous trees—trees that shed their leaves each year. These biomes usually have rich, moist soil and about 40 inches of rain each year. In spring and summer, tree leaves shade the forest floor. The deciduous forest is a friendly environment for many types of plants and animals, from earthworms and insects to songbirds, deer, foxes, and more.

Tundra

Tundra biomes are in far northern climates too cold for trees to grow. This region lies on top of a layer of ice that never melts, called ***permafrost***. Not many animals stay for winter in this harsh biome. Lemmings, arctic foxes, wolves, and polar bears are among the few that remain. For a few short months, low-growing tundra plants and flowers bloom. Birds and insects fly about, and herds of moose and caribou come north from the boreal forests to graze.

Coniferous Forests

Coniferous forests are made up mostly of cone-bearing trees (see Gymnosperms, pp. 20-21). They grow better in cooler, drier conditions than do deciduous trees. Coniferous forests are divided into two types—***temperate*** forests and ***boreal*** forests. Temperate forests grow in mild climates, usually along coastlines. The giant sequoia trees in the northwest United States grow in temperate forests. Boreal forests cover the northern regions of the world, where summer is short and winter is long. Pine, spruce, hemlock, and fir trees thrive in the boreal forest.

Moose, elk, and deer are among the animals that wander through. Migratory birds (see Migration, p. 39) nest and rear young here before traveling south as winter approaches.

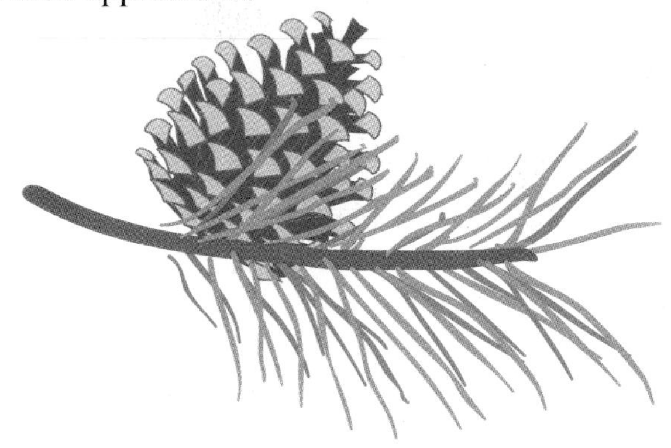

Tropical Rain Forests

Tropical rain forests are warm, wet biomes near or at the equator. They are filled with tens of thousands of different kinds of plants and animals. More species inhabit tropical rain forests than all other biomes combined.

The temperature stays around 80°F both day and night, all year long. Annual rainfall averages 100 to 200 inches and falls throughout the year. Rain forests are sometimes called ***jungles***. The ground, or ***floor***, of the forest is in almost total shade. A layer of shrubs and trees called the ***understory*** grows from about ten to fifty feet above the floor. The treetop layer above the understory is the ***canopy***, which is seventy-five to 100 feet above the floor.

Many plants and animals live in the canopy. Here flowers find sunlight, and ***arboreal***, or tree-dwelling, animals make their homes. Uncounted species of invertebrates, amphibians, reptiles, and birds live in tropical rain forests, along with monkeys, leopards, and other mammals.

Tropical Rain Forest in the News

Rain forests cover only two percent of the earth's surface, but more than half of all plants and animals live there. The largest tropical rain forest surrounds the Amazon River in South America. The human population of the region is growing fast. To make room for farming, ranching, and other activities, people cut or burn down the forest. Some experts estimate that 100 acres of rain forest are destroyed each minute.

Many people are worried about the loss of rain forests throughout the world. More than half of the world's rain forests have been chopped down in the last thirty years. This destruction has led to the loss of countless species of animals and plants. They have either died out or are in danger of becoming ***extinct*** (see p. 40). If the tropical rain forests were to be destroyed, the ***balance of nature*** (see p. 37) of the whole world might change. That could mean big trouble for humans as well as other plant and animal life!

Wake Up Call: Diurnal and Nocturnal Animals

Diurnal animals are active during the day and sleep at night. ***Nocturnal*** animals are active during the night and sleep during the day.

Most animals are diurnal. It's easier to see in daylight than in the dark, so it's easier to find food. But it's also easier to be seen, and many ***herbivores*** are easy prey during the daylight hours. So mice and other prey animals look for food at night. Some predators, such as owls and leopards, have ***adapted*** (see p. 38) to nocturnal life in order to prey on night-roaming creatures.

Still other animals, such as desert dwellers, are forced by climate to remain at rest during the day and avoid wandering about under the scorching sun.

Mice search for food at night. Owls have adapted to nocturnal hunting to prey on mice.

Ecology

Plants and animals live together and rely on each other in the world's ***biomes*** (see pp. 30-33). For example, animals eat plants and each other to get the food and nutrients that keep life going. The study of how plants and animals live together and interact with each other in their natural surroundings is called ***ecology***.

The natural home where a plant or animal finds the food, water, and space it needs to survive is called a ***habitat***. The group of different plants and animals that live in a habitat is a ***community***. The word ***ecosystem*** is used to describe how the plants and animals within a habitat interact with each other and with the nonliving parts of their environment.

Ecosystems

In an ***ecosystem***, plants and animals interact with each other and with their environment. ***Food chains*** and ***natural cycles*** are examples of two ways plants and animals interact with each other.

Food Chains and Food Webs

A ***food chain*** is a chain of living things. Simple food chains exist in your own backyard. The grass uses the sun's energy to make food (see Photosynthesis, p. 28). A sparrow eats grass seed from the lawn. Then your neighbor's cat eats the sparrow.

Producers
Green plants capture energy from the sun. Through the process of *photosynthesis*, they *produce* food.

Consumers
Herbivores, or plant-eating animals, *consume*, or eat, green plants. *Carnivores* then eat the plant-eating animals. Some carnivores feed on other meat eaters.

Decomposers
Bacteria and fungi break down dead plants and animals and use them for food. The breakdown process is called *decomposition*. Decomposition frees nutrients from the dead plant or animal. Plants take up and use the nutrients to grow.

Food webs are networks of food chains. Look again in your own backyard. The grass seed that feeds sparrows also feeds field mice. The neighbor's cat likes field mice, as well as sparrows, for dinner. Hawks and owls also catch mice. The simple food chain from grass seed to cat joins up with other food chains to form a network, or food web.

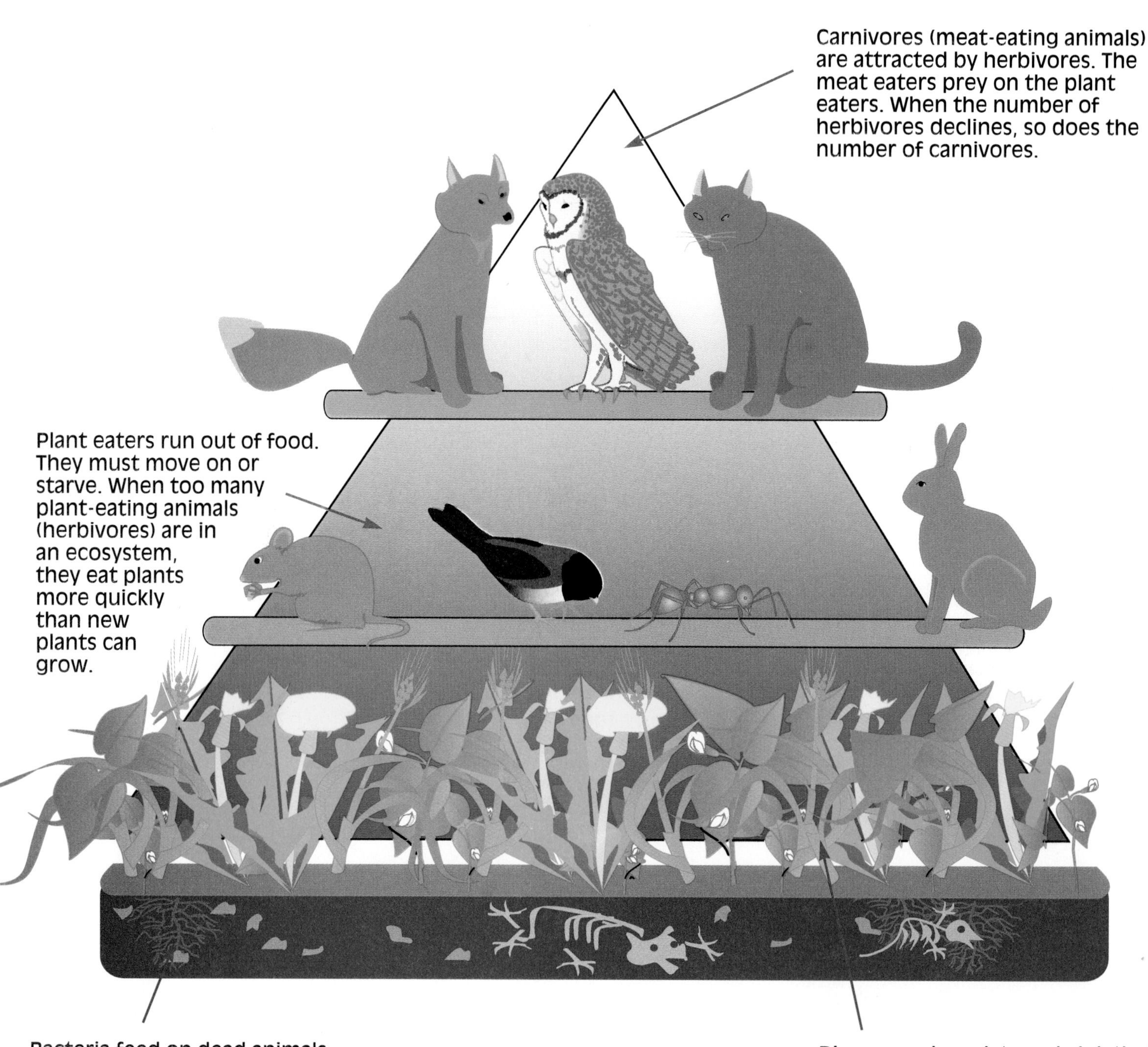

Carnivores (meat-eating animals) are attracted by herbivores. The meat eaters prey on the plant eaters. When the number of herbivores declines, so does the number of carnivores.
Plant eaters run out of food. They must move on or starve. When too many plant-eating animals (herbivores) are in an ecosystem, they eat plants more quickly than new plants can grow.
Bacteria feed on dead animals. When large numbers of animals die, big increases in bacteria may cause diseases in the plants and animals. Disease kills weak plants and animals.
Disease and predators shrink the herbivore population. Plants grow back. Balance returns to the ecosystem.

Natural Cycles

Water and certain chemicals—such as oxygen, carbon, and nitrogen—are constantly being exchanged between air, water, soil, plants, and animals.

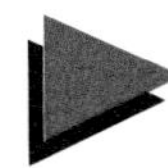

The number of plants and animals in an ecosystem must stay in balance to survive. Competing for food, water, light, and other resources is one way plants and animals stay in balance. This balance is called **homeostasis.**

Water Cycle

The sun heats water in soil, rivers, lakes and oceans, causing it to evaporate and become water vapor, a gas.

Water vapor rises, cools, and condenses to form tiny water droplets or ice crystals in clouds. The water falls back to earth as rain, snow, or other precipitation.

Most water returns to the sea or sinks into underground water sources.

Plants take up water from the soil through their roots. They *transpire* (see p. 29) most of the water they take up.

Water makes up almost seventy-five percent of living things. When plants *decompose*, water is released.

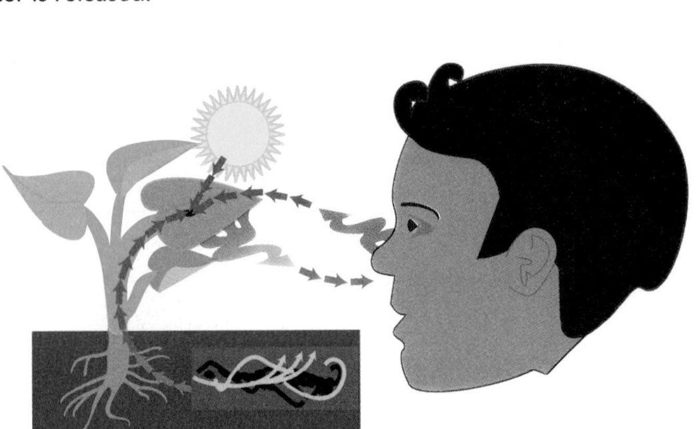

Carbon and Oxygen Cycle

Carbon and oxygen make up the gas *carbon dioxide*. Plants take in carbon dioxide during *photosynthesis*. Plants use the carbon from carbon dioxide to make food during photosynthesis. This food provides matter and energy to form new plant cells.

During photosynthesis, plants also release oxygen into the air. Animals breathe in oxygen. Plants also take in oxygen during *respiration* (see p. 29). Both animals and plants give off carbon dioxide during respiration.

Dead plants and animals produce carbon dioxide as they decay.

Carbon from ancient dead plants is stored as *fossil fuels*, such as coal, oil, and gas.

Nitrogen Cycle

Nitrogen gas makes up about seventy-eight percent of our air. Nitrogen is an important part of proteins and other plant and animal matter. But plants and animals cannot use nitrogen directly from the air.

Instead:

1. Certain bacteria in the soil and plant roots are able to take nitrogen gas from the air and change it into a form of nitrogen that plant roots can take up and use. This process is called *nitrogen fixation*.
2. Some tissue in dead plants and animals, and even in animal wastes, contains nitrogen. Several kinds of soil bacteria break down the nitrogen-containing tissues and change them into *nitrates*, which plants absorb through their roots. The cycle is complete when other soil bacteria take in nitrates and release nitrogen gas back into the air.

Plant and Animal Behavior

In order to live within their ecosystems, plants and animals behave in special ways. Behavior in ecology refers to how animals eat, sleep, communicate, and reproduce (see pp.18-19, 34). In plants, behavior might refer to how plants reproduce and make food (see p. 28 and pp. 25-27). Behavior in ecology also means how plants and animals cope with changes in their environments.

Adaptation

Sometimes conditions in an ecosystem change. Droughts, floods, temperatures, and rainfall affect ecosystems. People and animals destroy habitats by overhunting, over-grazing, and polluting (see p. 40).

Changes in an ecosystem always affect its plants and animals—including humans! Some changes may be so harmful that some plants and animals are destroyed. Other changes may be so big that only plants and animals that ***adapt***, or change over time, survive. The gradual change of plants and animals to suit their environments is called ***adaptation***.

Camouflage

Arctic foxes and chameleons are two of the many animals that blend in with their surroundings by ***camouflage***. Camouflage protects animals from being seen by their enemies and allows them to sneak up more easily on their prey.

Chameleon's skin can change into a remarkable variety of colors, from dark browns to bright greens. The change takes only seconds, so as a chameleon moves, it often changes colors to blend with its environment.

In summer months, the Arctic fox has a brown coat. In winter months, the brown coat is replaced with a white coat. In these two extreme seasons, the Arctic fox blends in with its surroundings.

Arctic fox

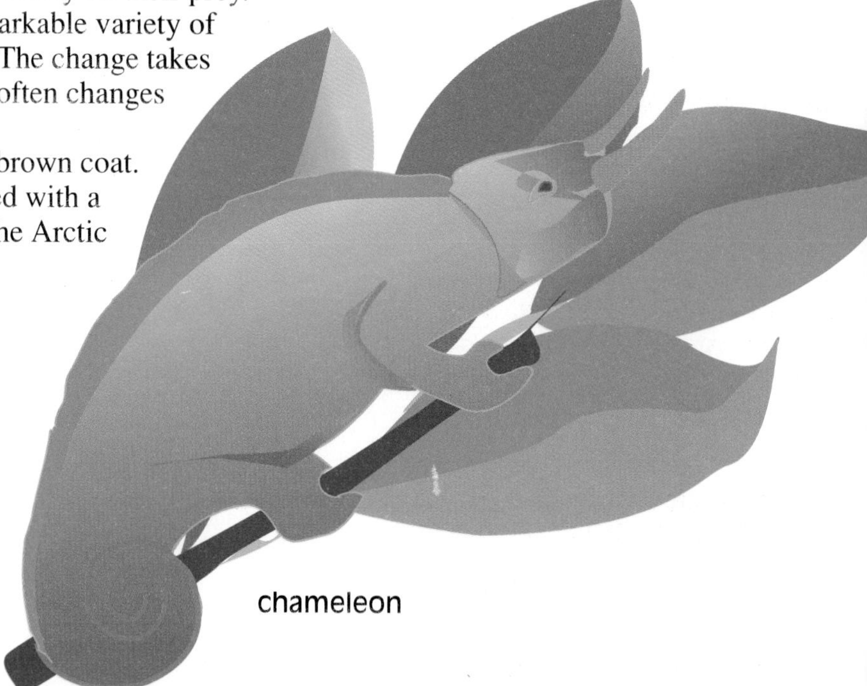

chameleon

Migration

Migration is the movement of animals over the same route in the same season each year. Animals migrate to find food. They wander from a winter place where food is scarce to a warmer place where more plants and animals live. Fish usually migrate to breed.

Migration by Air

Animal	Distance	To and From
Arctic Tern	22,500 miles round trip	North Pole to South Pole
Bat	2,400 miles round trip	Canada to Bermuda (in Atlantic Ocean)
Golden Plover	19,000 miles round trip	Alaska and Canada to South America
Monarch Butterfly	2,000 miles round trip	U.S. and Canada to Mexico
Short-tailed Shearwater	20,000 miles round trip	Australia to Alaska and Northern Pacific Ocean
Wandering Albatross	18,000 miles	Circles the entire globe, west to east

Migration at Sea

Animal	Distance	To and From
Atlantic Salmon	6,000 miles round trip	St. Lawrence River to Atlantic Ocean
Blue Whale	5,000 miles round trip	Indian Ocean to Antarctica
Dogfish	2,500 miles round trip	Coast of Canada to Mediterranean Sea
Emperor Penguin	Up to 2,000 miles yearly	Across Antarctica
Fur Seal	9,000 miles round trip	Pribilof Islands, Alaska, to coast of California
Green Turtle	3,000 miles round trip	South America to Ascension Island (Atlantic Ocean)
Humpback Whale	8,000 miles round trip	Indian Ocean to Atlantic Ocean

Migration by Land

Animal	Distance	To and From
Bison	800 miles	Within Northern Canada
Caribou	800 miles	Alaska and Canada to Arctic Circle
Wildebeest	900 miles	Serengeti north into Kenya (Africa)

Hibernation

Some animals survive seasonal changes in food supply by migration. Others enter a deep sleep called ***hibernation*** to wait until the food in their ecosystem is plentiful again. During hibernation, the animals' body processes slow down, and they survive on stored food or fat.

Many different animals hibernate—insects, birds, reptiles, amphibians, and mammals, from tiny rodents to huge bears. These animals eat heavily in the autumn months to store up fat. Then they burrow into the ground, curl up under leaves, or hide themselves in dens, safe from the winter cold and enemies.

During hibernation

1. **Heartbeats, breathing, and other bodily functions slow down.**
2. **The body hardly moves.**
3. **The body burns its stored fat to survive.**

3 Environmental Issues and Conservation

To *conserve* means to "protect" or "take care of." In ecology, conservation means to protect the earth's plants, animals, habitats, and natural resources. But from what does the earth need to be protected?

Too Many People

The number of people on the earth is growing faster than the supplies of food and water. This creates a problem called ***overpopulation***. If a place is overpopulated, there are not enough resources for people to survive. These same resources are also needed by other animals and plants. Many animals have become ***extinct***, or died out completely. Others are threatened with extinction, or are ***endangered***.

Extinct and Endangered Species

Many animals that once lived on the earth no longer survive. They are ***extinct***. Sometimes animals die out from natural causes or catastrophies, for example, the dinosaurs (see pp. 16-17). Others die out because humans have hunted them down or destroyed or polluted their habitats. Some animals are in danger of becoming extinct. Some have lost their habitats to people or pollution. Others have been hunted down to very small numbers. These animals are ***endangered***.

In 1973, the United States government passed the Endangered Species Act. The law is meant to protect endangered animals and help them survive. The act has saved many species, including our national symbol, the bald eagle.

ENDANGERED SPECIES

Common name	Scientific name	Range
Mammals		
Bobcat	*Felis rufus escuinapae*	Central Mexico
Brown or grizzly bear	*Ursus arctos horribilis*	U.S.
Cheetah	*Acinonoyx jubatus*	Africa to India
Ozark big-eared bat	*Plecotus townsendit*	Midwest, Southwest U.S.
Birds		
American peregrine falcon	*Falco peregrinus anatum*	Canada to Mexico
Bald eagle	*Haliaeetus leucocephalus*	U.S. (most states), Canada
California condor	*Gymnogyps californianus*	Western U.S.
Hooded crane	*Grus monacha*	Japan, Siberia

Pollution

Pollution is the damage done to the environment by harmful substances. Pollution comes in many forms, from litter and noise to ***acid rain*** and oil spills. Pollution affects the air, land, and water, and may even be changing the outer layers of the earth's atmosphere (see p. 44).

Garbage

Garbage is ***solid waste***. It includes candy wrappers, old newspapers, soda cans, milk cartons, juice bottles, used diapers, plastic detergent bottles, half-eaten sandwiches, apple cores, and so on. Solid waste often winds up in landfills at the dump. Garbage is ***litter*** when it is dumped carelessly where it doesn't belong. Some garbage, such as fruit and vegetable peels, decays quickly and actually nourishes the soil when composted. But plastics, glass, and metals do not decay very fast. Many of these materials should be ***recycled*** or ***incinerated*** (burned).

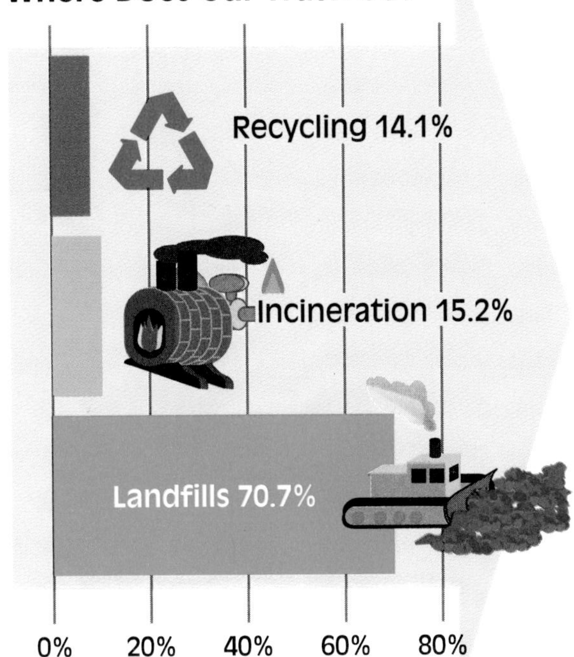

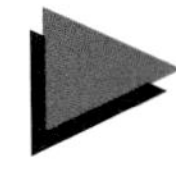

Recycling ***helps conserve natural resources by creating less garbage and by breaking down used products to make other things from them. Today we recycle aluminum cans and other metals, glass jars and bottles, paper, some plastics, and other materials. Other conservationists try to reuse resources—such as glass bottles and jars—without changing them at all.***

Some Conservation Stats You Should Know

- About twenty-seven million more acres of trees are cut down each year than are planted.
- Every day American families produce an estimated four million pounds of dangerous household wastes (paints, paint thinners, car batteries, and so on).
- Petroleum (oil-based products) is the major source of energy in the United States, followed by natural gas and coal. Nuclear energy and solar and water power account for only fifteen percent of the energy supply.
- Americans use more energy than they produce with their own resources. Almost half the petroleum products used in America are imported (brought in) from other countries.
- One edition of a major daily newspaper, such as the *New York Times*, the *Chicago Tribune*, or the *Los Angeles Times*, uses wood from as many as 5,000 trees.
- In Europe and North America, most people each use about two trees' worth of paper in one year.
- Paper makes up about forty-one percent of the trash in the United States.
- Each ton of recycled paper saves more than three cubic yards of landfill space and 380 gallons of oil.

Noise Pollution

Machines, traffic, airplanes, boom boxes, car horns, alarms, and sirens make loud sounds. These sounds can add up to noise pollution. Noise pollution can make people grouchy and can harm their ability to hear.

Air Pollution

Cars, factories, and power plants pump gases, smoke, soot, and chemicals into the air. These pollutants are left over when ***fossil fuels*** are burned to create energy. Polluted air is harmful to breathe, not only for people, but also for other animals and plants. Air pollution has been linked to illness, brain damage, and death.

Some pollutants in the air fall back to the earth in ***acid rain***. Scientists also fear that air pollution might add to ***global warming*** (see p. 44).

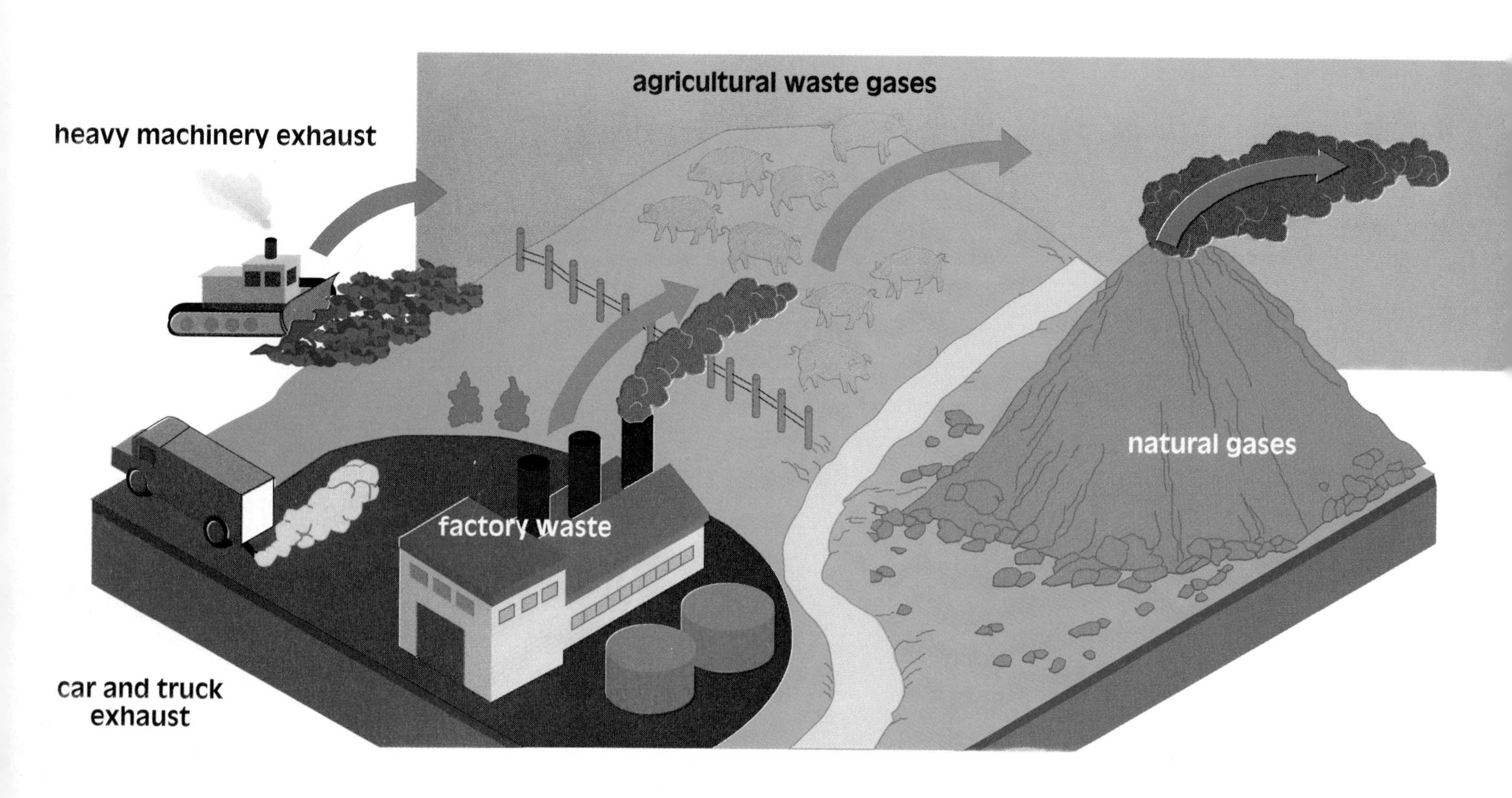

OZONE DESTRUCTION

The ***ozone layer*** is a thin layer of gas in the atmosphere. It is wearing away because chemicals called ***chlorofluoro-carbons*** are released into the air from refrigerators, air conditioners, and aerosol cans.

Since the ozone layer filters out harmful ultraviolet rays from the sun, the destruction of the ozone layer also threatens the plants and animals on the earth's surface.

Water and Land Pollution

Factories sometimes dump solid waste into landfills and into lakes, rivers, and seas. Some towns and cities dump untreated human sewage into the water, too. These pollutants kill fish and underwater plant life. Fertilizers and pesticides used on farms and to conrol weeds also wash into waterways, polluting the water and endangering the life in lakes, rivers, and oceans. The pollution can also seep into the water under the ground.

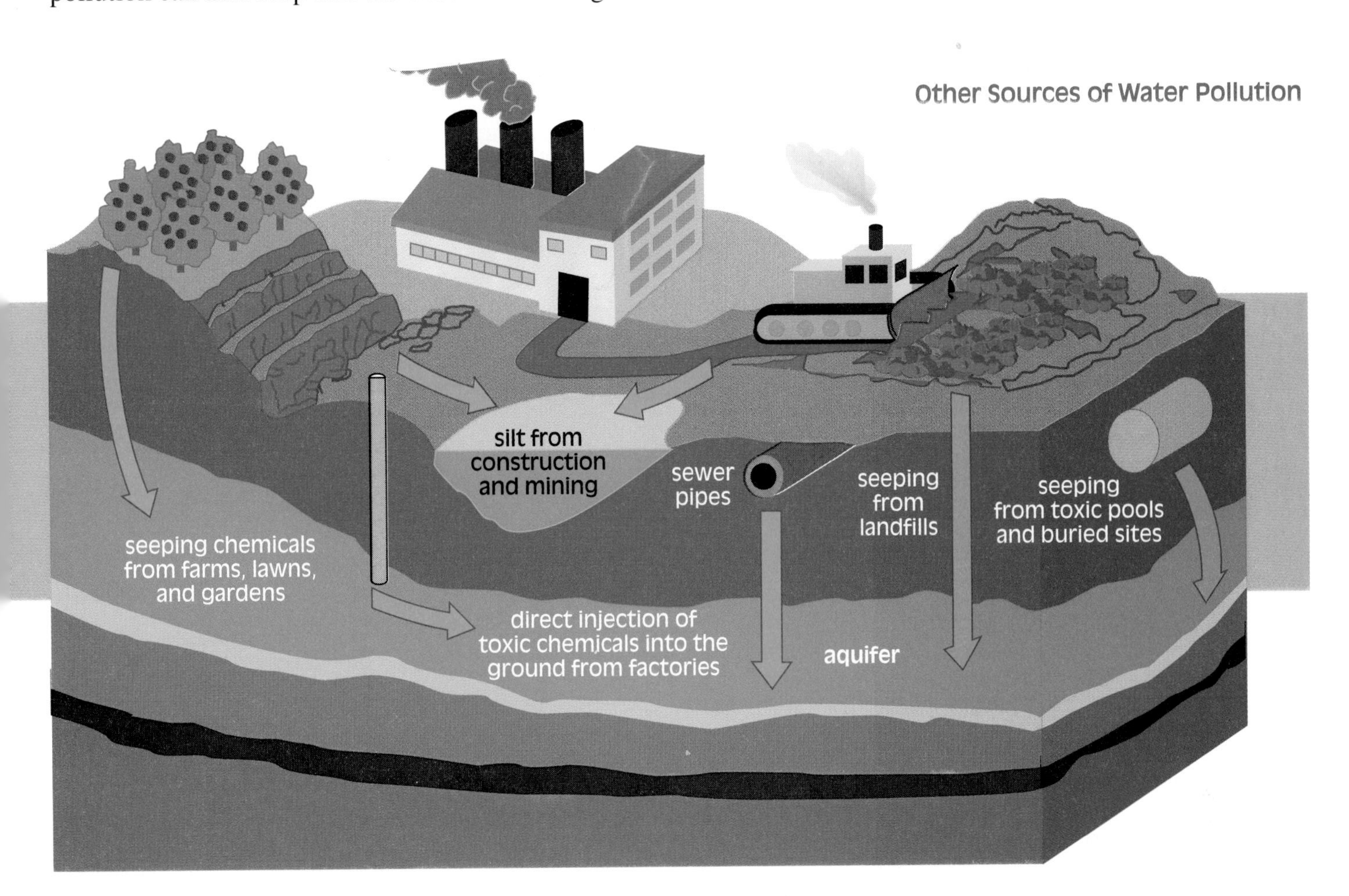

Thermal Pollution and Radioactive Waste

Nuclear power is an alternative to burning ***fossil fuels*** to make energy (see p. 101). Although nuclear power plants do not pollute the air, they do produce solid ***radioactive*** wastes, which can cause illness and death.

Nuclear power plants, like other power plants, also produce heat, or ***thermal***, pollution. The power plants are cooled with water from nearby rivers, lakes, and seas. When this water returns to the rivers, lakes, and seas, it is hotter than it was before. This heated water may harm plants and animals.

Global Warming and the Greenhouse Effect

The earth's atmosphere keeps the planet warm by a process called the ***greenhouse effect***. It lets in the sun's energy in the form of visible light. When this light hits the earth, it becomes heat. Greenhouse gases let some of this heat back into space, but they trap some of it within the atmosphere. Burning fossil fuels, such as coal and gasoline, adds more greenhouse gases, such as carbon dioxide, to the atmosphere.

Some people think that the increase in the amount of greenhouse gases may lead to an increase in the world's temperature. They call this ***global warming***. They are concerned that this warming could harm life on earth. However, the earth's atmosphere and climate are complicated, and no one can be sure how much it might warm up and how much harm that could do.

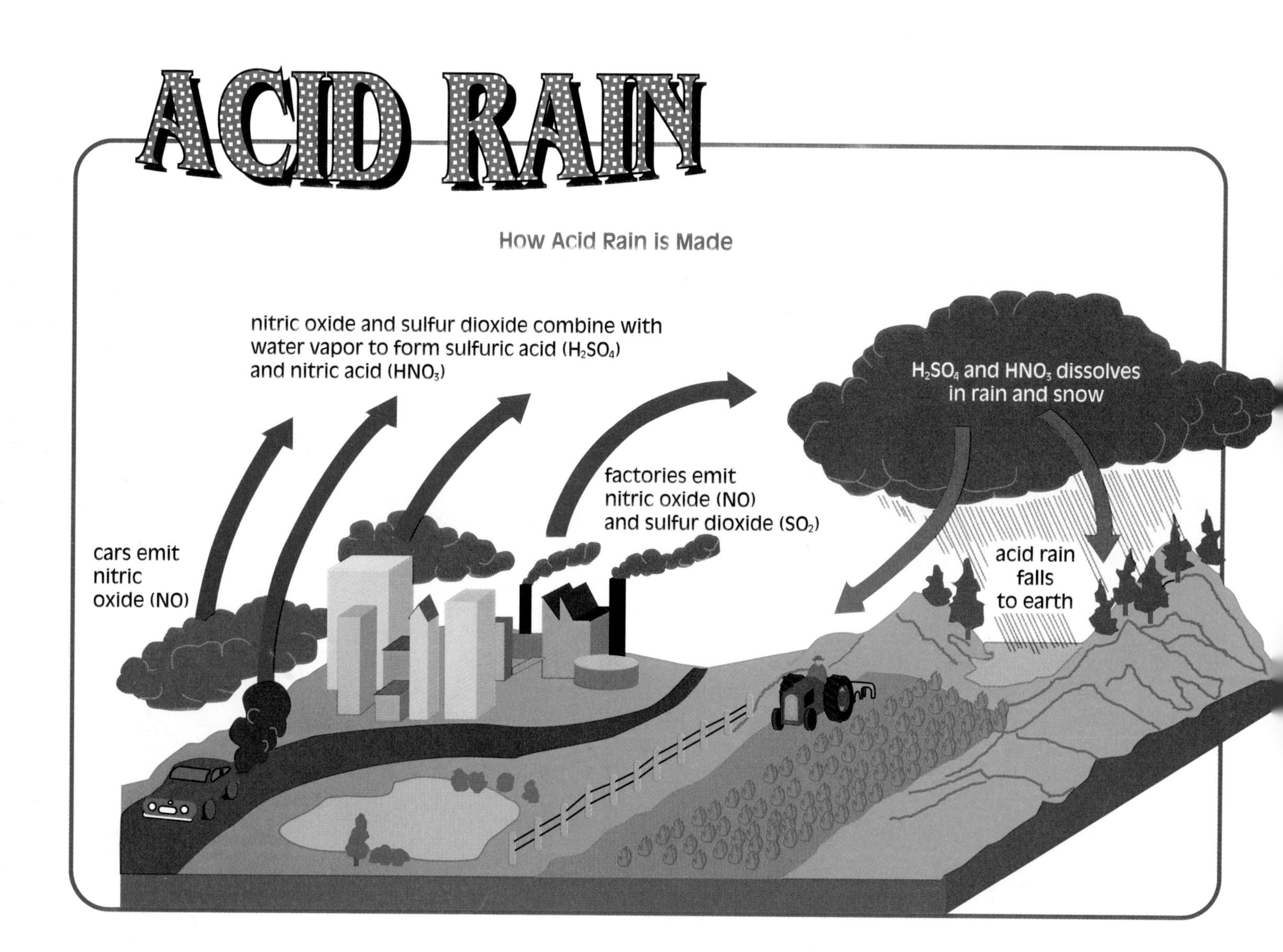
ACID RAIN
How Acid Rain is Made
nitric oxide and sulfur dioxide combine with water vapor to form sulfuric acid (H_2SO_4) and nitric acid (HNO_3)
H_2SO_4 and HNO_3 dissolves in rain and snow
factories emit nitric oxide (NO) and sulfur dioxide (SO_2)
cars emit nitric oxide (NO)
acid rain falls to earth

CONSERVATION Begins at Home

- Help your parents set up a carpool for trips to school, sports practices, and after school activities. One car uses less gas and creates less exhaust than several cars.
- Use rechargeable batteries instead of the kind you throw away.
- Reuse brown paper bags for garbage bags, book covers, package wrap, and so on.

- Turn off the lights when you leave a room and unplug electric toys, games, hair dryers, and other items when you're done using them.
- Pack your lunch in reusable containers. Avoid using plastic utensils and sandwich bags.
- Donate old toys and clothes to charities rather than throw them away.

- Recycle newpapers and magazines, metal cans, glass bottles, and plastic containers. If your town doesn't have a pick-up service for recyclables, find out how you can recycle.
- Use both sides of a piece of paper before throwing it away.
- Ask your parents to use reusable shopping bags for groceries.

- Don't leave the water running when you're brushing your teeth or doing the dishes. Run the water only as needed to wash and rinse.
- Take a shower instead of a bath to save water. Also, install a water-saving shower head in your bathroom.

WELCOME

PLANET EARTH

1 Earth, Then and Now

Scientists believe that the earth formed from a cloud of dust particles about 4.6 billion years ago. The dust cloud came together to form a ball of ***molten***, or melted, rock. The ball cooled over several million years, forming a crust on the outside.

The surface of the earth is the ***crust.*** It is a thin layer of rock. It ranges from three miles thick under the deepest parts of the oceans to twenty-two miles thick under mountain peaks.

Beneath the crust is the ***mantle***, a layer of very hot, sometimes molten rock about 1,800 miles thick.

At the center of the earth is the ***core.*** The core is made up of two parts: the ***outer*** core, which is made up of molten rock; and the solid ***inner*** core. The core has a radius of about 2,100 miles.

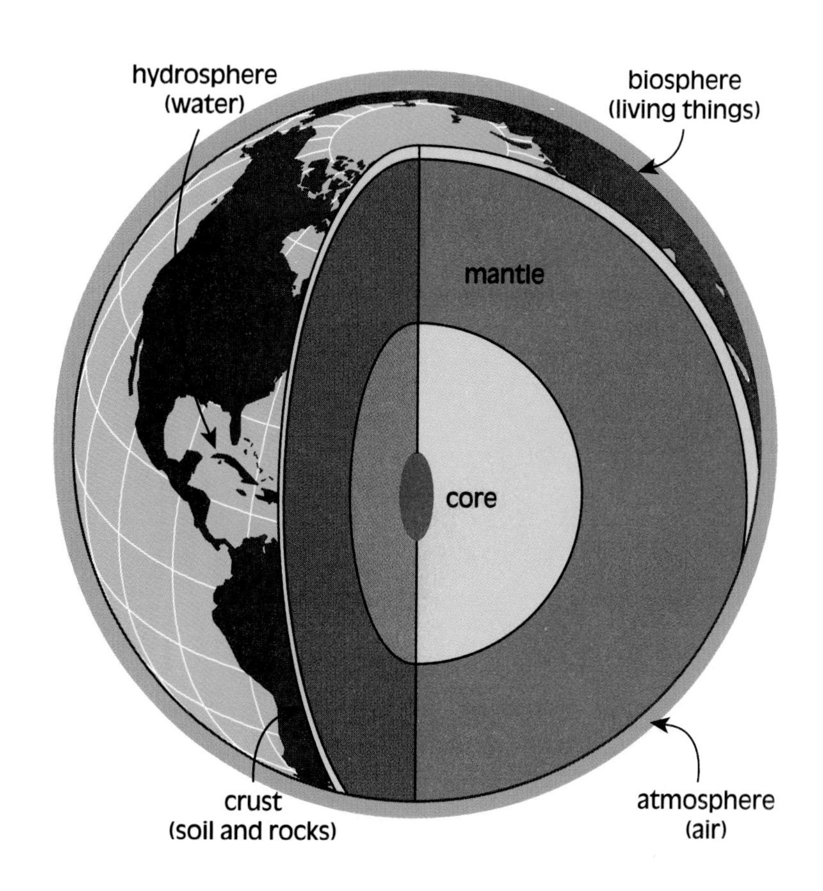

MYA*	2		65	144	208	248
PERIOD	QUARTERNARY	TERTIARY	CRETACEOUS	JURASSIC	TRIASSIC	PERMIA
ERA	CENOZOIC		MESOZOIC			

* MYA means *million years ago*

Dating the Earth

When geologists first began studying the earth, they knew it was very, very old. But they weren't sure how old.

Then geologists took a closer look at rock formations. They found many **fossils**, the remains of ancient plants and animals, in sedimentary rocks. Sedimentary rocks are formed in layers, with the oldest layers at the bottom. Fossils within these layers show the plants and animals that existed when the rocks formed. So, geologists could figure out which parts of the earth were the oldest based on the ages of fossils in the rock layers. This method is called ***relative dating***.

In the 1900s, scientists began to understand ***radioactivity***. If something is ***radioactive***, it breaks down, or ***decays***, at a constant rate, and scientists know this rate. For example, scientists know that the element uranium (see pp. 98–99) breaks down into lead over a certain amount of time. By measuring the amount of uranium and lead in a rock, geologists can measure the age of the rock. They know how long it takes for the lead to form. This method is called ***absolute dating***.

Plate Tectonics

The surface, or crust, of the earth is always moving. The continents drift apart, the oceans rise and fall. Over time, the earth's movements have caused the thin surface rock to break up into ***tectonic plates***. The plates are like enormous rafts, floating upon the earth's mantle.

Continental drift (see p. 52) is caused by the movement of these plates. The plate movement also forms mountains and volcanoes (see p. 58).

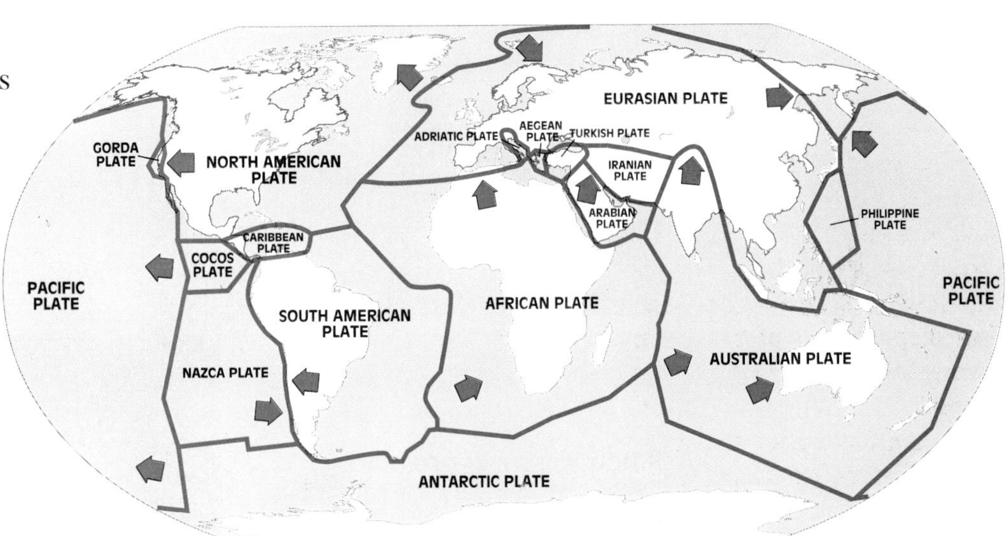

Eras of Geologic Time

About 3.5 billion years ago, life appeared on the earth. How these life forms came to be is a mystery, but scientists think that chemicals from volcanic eruptions became part of the atmosphere. These chemicals include hydrogen compounds, methane, ammonia, and water. Many scientists think that these chemicals came together to form ***amino acids***. Amino acids are the building blocks of ***proteins*** and simple cells (see p. 6).

320	360	408	438	505	550	4600
YLVANIAN America)	MISSISSIPPIAN (North America)	DEVONIAN	SILURIAN	ORDOVICIAN	CAMBRIAN	PRECAMBRIAN TIME
CARBONIFEROUS						
PALEOZOIC						

2 The Land

Rocks, Minerals, and Soil

Rocks

Rocks are grouped into three types—***igneous***, ***sedimentary***, and ***metamorphic***.

Volcanoes bring red hot *magma* (molten rock) out of the earth's core. When magma cools, it forms igneous rock.

Sedimentary rocks are made of *sediment*, tiny grains of sand and bits of other rocks. The sediment settles at the bottom of lakes, rivers, or oceans, and slowly hardens over time. The rocks "grow" as more sediment settles and become layered.

Metamorphic rocks form from sedimentary or igneous rocks that have *metamorphosed*, or changed, over time. Heat and pressure work together to harden the minerals in the sedimentary and igneous rocks, finally changing them into a new, metamorphic rocks.

THE ROCK CYCLE

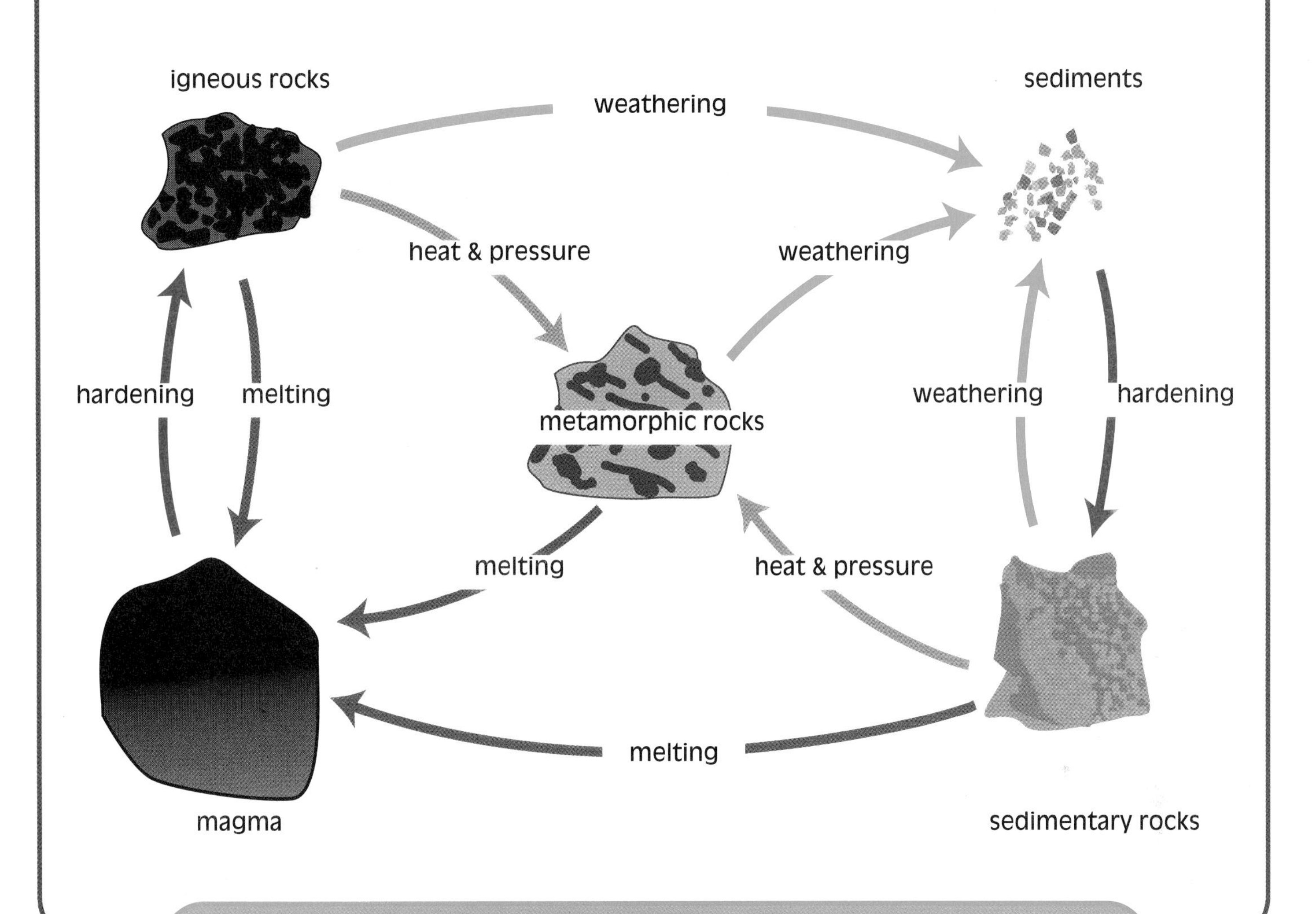

Common Rocks

sedimentary	metamorphic	igneous
sandstone	slate	granite
limestone	marble	obsidian
shale	quartzite	basalt

Minerals

Scientists have identified more than 2,500 different ***minerals***. Minerals are natural, non-living substances made up of particles that are arranged in regular patterns. They are the building blocks of rocks. Minerals include salt, talc, clay, metal ores, and gemstones. Some rocks consist of only two minerals, while others are made up of hundreds. Minerals give rocks their basic qualities—color, hardness, and ***luster.*** (Luster means that the rock looks metallic, or glassy, or pearly, and so on.)

As magma cools, chemicals in the magma form minerals. As igneous rocks change into metamorphic rocks, the minerals change, too, into new forms of minerals. Some minerals are dissolved in underground water. Over time, these minerals harden into crystals.

When many minerals are located in the same area, they can be mined. Minerals that are mined are called ***ore***.

The Measure of a Mineral

Minerals are identified by hardness, based on a scale developed by the German mineralogist Friedrich Mohs in 1822. The Mohs scale ranks minerals on a scale of one to ten. One is the softest mineral, talc. Ten is the hardest mineral, diamond. A mineral can scratch a mineral softer than itself, but not a mineral harder than itself. Mohs scale rates the hardness of minerals using common minerals of different hardness.

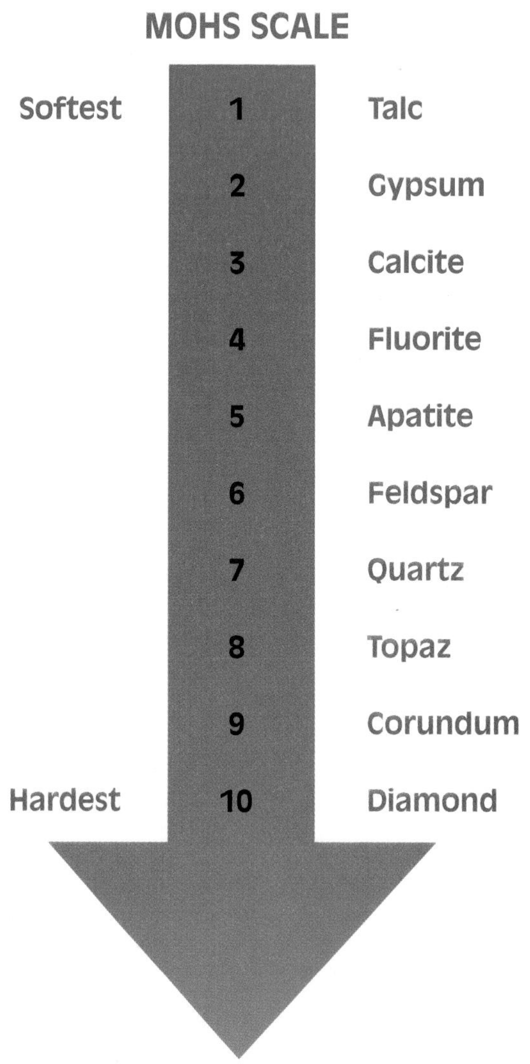

MOHS SCALE

Softest	1	Talc
	2	Gypsum
	3	Calcite
	4	Fluorite
	5	Apatite
	6	Feldspar
	7	Quartz
	8	Topaz
	9	Corundum
Hardest	10	Diamond

Soil

Soil is a combination of ***organic*** (once-living) particles called ***humus*** and ***nonorganic*** particles (bits of rocks and minerals). Soil is located between the atmosphere and rocky layers of the earth. The thickness of soil changes from place to place. It can be several feet or only a fraction of an inch thick.

Soil is divided into four types, based on the size of the rock particles it contains.

Four Types Of Soil

Clay	**Very fine particles, less than a ten-thousandth of an inch thick.**
Silt	**Fine particles, ranging from a ten-thousandth to two-thousandths of an inch thick.**
Sand	**Coarser particles, ranging from two-thousandths to eight-hundredths of an inch thick.**
Loam	**A mixture of clay, sand, silt, and organic matter.**

Land

Almost one-third of the earth's surface is covered by ***continents***. Scientifically, there are six: Eurasia, Africa, Australia, North America, South America, and Antarctica. (Geographers separate Eurasia into Europe and Asia.) Continents are large masses of land made up of a central slab of rock covered in sedimentary rock (see p. 48), often bordered on the edges by mountains. The land on most continents is about twenty-five miles thick.

The rest of the land on the earth's surface is made up of large islands and island groups that rise above the oceans and seas. Most of these islands were formed by ***volcanoes*** (see p. 58).

Continent	Area (million square miles)
North America	**9.4**
South America	**6.9**
Eurasia	**21.2**
Africa	**11.7**
Australia	**3.3**
Antartica	**5.4**

PANGAEA: GET MY DRIFT?

About 200 million years ago, all the continents were connected. They formed a ***supercontinent*** that scientists call Pangaea. Scientists studied fossils and rock types to help prove the connections.

Pangaea broke up, and the continents moved apart to form the continents we know today.

In another 200 million years, the earth may look completely different again. That's because the continents are still moving. The movement is called ***continental drift*** (see p. 47).

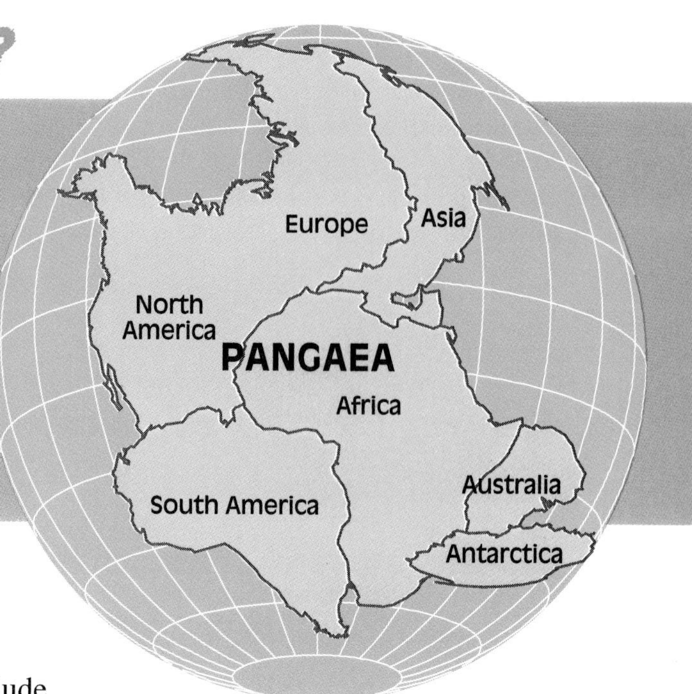

Landforms

On the land are several different ***landforms.*** Landforms include mountains, hills, valleys, plains, and plateaus.

Most landforms are created through natural processes, such as ***erosion*** from wind and rain and ***weathering*** from frost and sunshine. Other natural events, such as ***volcanoes*** and ***earthquakes***, also shape the land.

Mountains

A mountain is any point that rises at least 1,000 feet above its surroundings. Some mountains are steep and snowcapped. Others have gentle slopes and rounded tops. Some, like volcanoes, have giant holes in their tops. We now know that there are also mountains under the oceans, taller even than Mount Everest, the tallest mountain on land.

How Mountains Are Formed

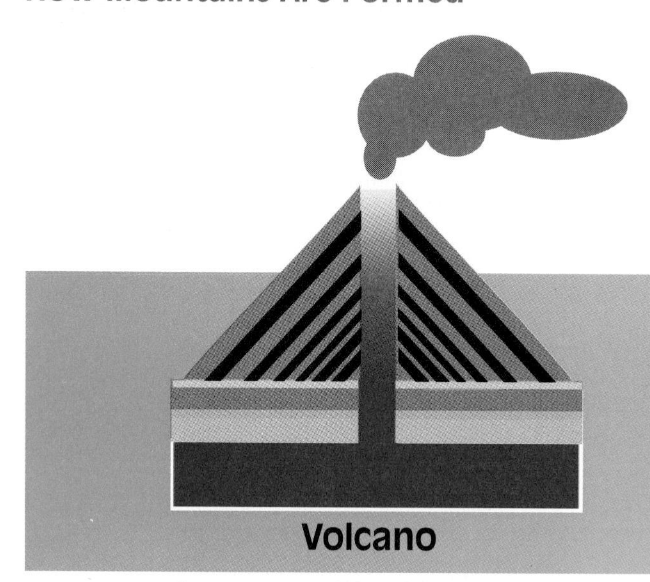

Some mountains form when molten rock from deep inside the earth rises to the surface, then pours out in the form of lava (see Volcanoes, p. 58).

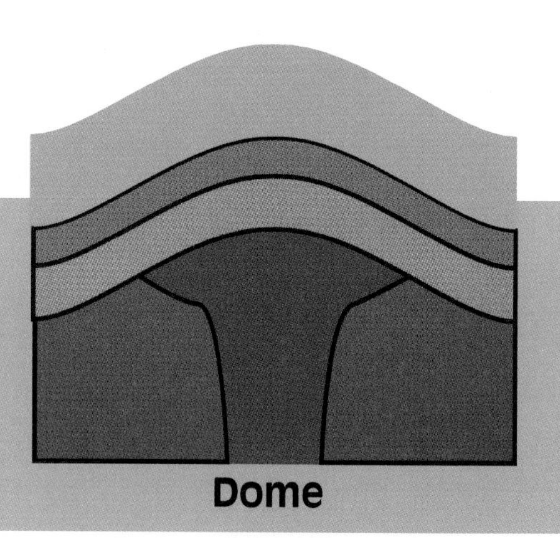

Sometimes molten lava pushes up toward the earth's surface, but does not break through.

Natural Rubdowns

EROSION AND WEATHERING

When water runs rapidly downhill, it carves gullies in the land, taking with it large amounts of dirt and rock. This process is called gully or rill erosion.

Water running over gentle slopes takes earth with it as it moves from a large area. This process is called sheet erosion.

Rivers and streams are the great movers of earth. When rivers run into lakes or oceans, they dump tons of dirt that they have carried for long distances. This dirt is called silt when it is deposited in standing water.

The wind blows dirt from the surface of the land when it is not protected with a covering of vegetation.

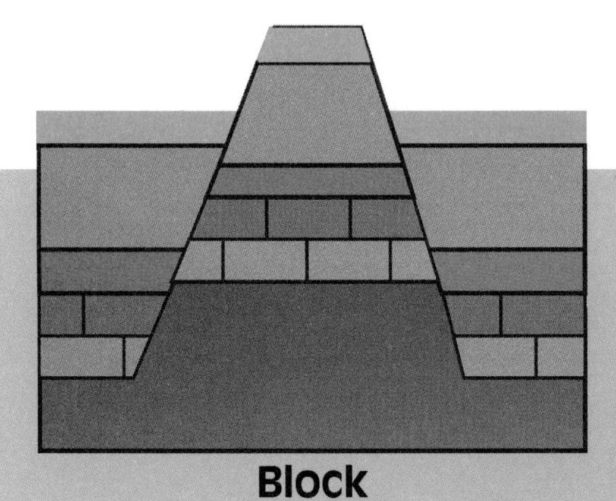

Blocks of rock split along fault lines and slide in opposite directions.

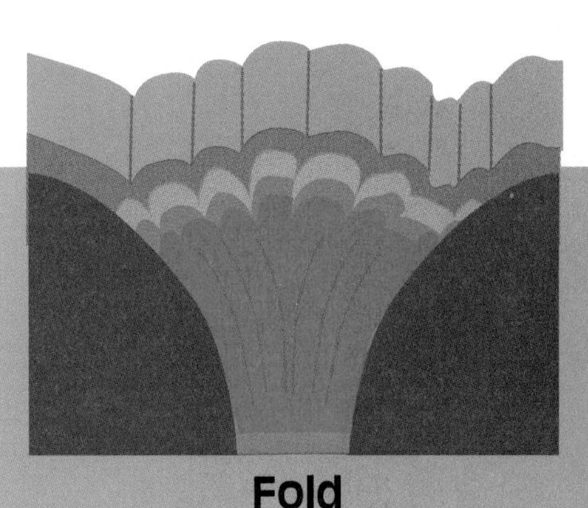

As the great plates on the earth's crust move toward each other, the sediments deep below the earth's surface are squeezed up to form mountain ranges.

Hills

Like mountains, hills are high points on the earth's surface, but they are not as high. Some hills are formed the same way as mountains. Others are formed by gigantic sheets of ice called ***glaciers*** (see p. 67).

Long ago, ice sheets covered huge portions of the earth's surface. These ice sheets moved back and forth over the land, somtimes getting bigger, sometimes shrinking. As they moved, bits of rock and soil, called ***glacial till***, were worn away from some areas and then dropped off on other areas. The glacial till built up to form different types of hills.

THE WORLD'S HIGHEST MOUNTAINS

Continent	Peak	Place	Height (feet)
North America	McKinley	Alaska	20,320
	Logan	Yukon	19,850
	Citlaltepec (Orizaba)	Mexico	18,700
	St. Elias	Alaska-Yukon	18,008
	Popocatepetl	Mexico	17,887
South America	Aconcagua	Argentina	22,834
	Ojos del Salado	Argentina-Chile	22,572
	Bonete	Argentina	22,546
	Tupungato	Argentina-Chile	22,310
	Pissis	Argentina	22,241
	Mercedario	Argentina	22,211
Africa	Kilimanjaro	Tanzania	19,340
	Kenya	Kenya	17,058
	Margherita PK.	Uganda-Zaire	16,763
	Ras Dashan	Ethiopia	15,158
	Meru	Tanzania	14,979
	Elgon	Kenya-Uganda	14,178
Southeast Asia, Australia, New Zealand			
	Trikora	New Guinea	16,685
	Jaja	New Guinea	16,500
	Mandala	New Guinea	15,420
	Wilhelm	New Guinea	14,793
	Kinabalu	Malaysia	13,455
Europe	Mont Blanc	France-Italy	15,771
	Monte Rosa	Switzerland	15,203
	Dom	Switzerland	14,911
	Liskkamm	Italy-Switzerland	14,852
	Weisshom	Switzerland	14,780
	Taschhom	Switzerland	14,733
Asia	Everest	Nepal-Tibet	29,028
	K2 (Godwin Austen)	Kashmir	28,250
	Kanchenjunga	India-Nepal	28,208
	Lhotse I (Everest)	Nepal-Tibet	27,923
Antarctica	Vinson Massif		16,864
	Tyree		16,290
	Shinn		15,750
	Gardner		15,375
	Epperly		15,100

Valleys

When water runs down mountain and hillsides, it wears away the rocks and soil it passes over. Over time, the water carves out V-shaped grooves. The grooves get deeper and wider, finally forming lowland areas called ***valleys***. Other valleys, typically U-shaped, are formed by glaciers (see p. 67).

Plains

Large areas of level or gently rolling land are called ***plains***. Plains can be dry and grassy farmlands or wet lowlands surrounding the mouths of rivers and streams. These plains are known as ***floodplains***.

Plateaus

Plateaus are areas of high, flat land. Plateaus are surrounded by steep rock faces called ***cliffs***. Because they rise straight up from the land, plateaus are often called ***tablelands***.

A **canyon** *is a valley with very steep sides. A* **ravine** *or a* **gorge** *is a small canyon. A* **gulley** *is a small ravine.*

How Valleys Are Formed

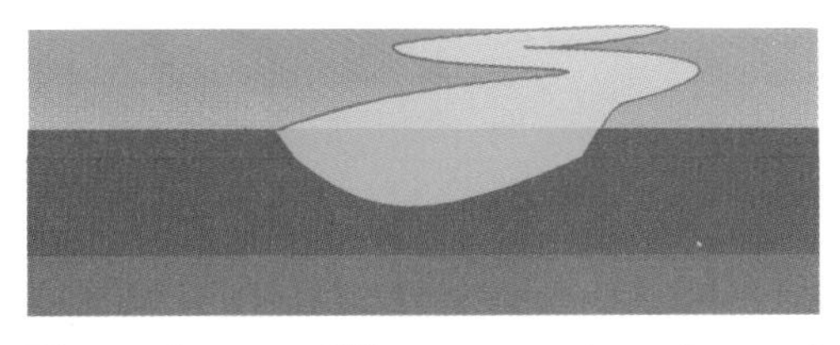

Rivers (see p. 65) carve out rocks and soil, too. The water evaporates and the river shrinks.

Glaciers (see p. 67) carve out rocks and soil. Over time, the glacial waters melt and evaporate, leaving a glacial valley.

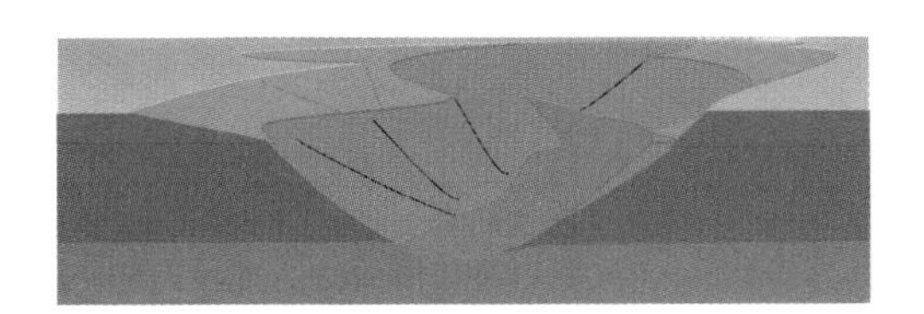

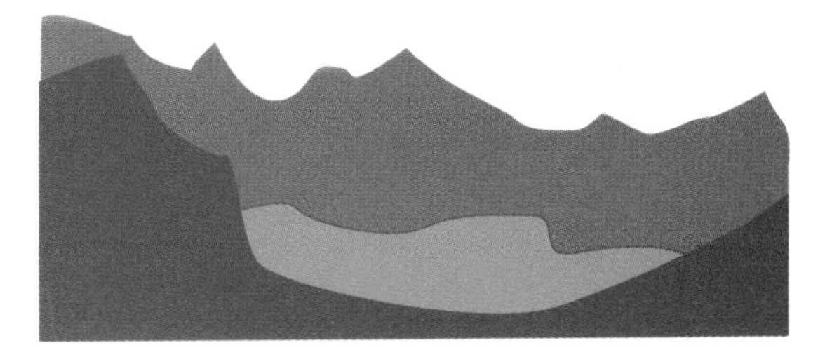

Prairies *and* **savannas** *are grassy plains without trees.*

Natural Events that Shape the Land

Earthquakes

Earthquakes are sudden, violent shifting movements in the earth's crust that can release more energy than a nuclear bomb. During earthquakes, the land trembles and shakes, often violently enough to topple buildings and collapse bridges. Sometimes the land cracks open.

Earthquakes are caused when the ***tectonic plates*** (see p. 47) that make up the earth's surface collide, separate, or scrape against each other. Because one plate can move faster or in a different direction than another, the plates do not always slip smoothly past each other. When that happens, the earth's surface jerks and shifts along the plate boundaries. These boundaries are called ***fault lines***. But not all fault lines are located on plate boundaries.

Forces push against blocks of crust on both sides of a fault. Friction prevents the blocks from sliding smoothly past each other.

Forces push over a period of thousands of years. The forces bend and twist the rocks along the fault.

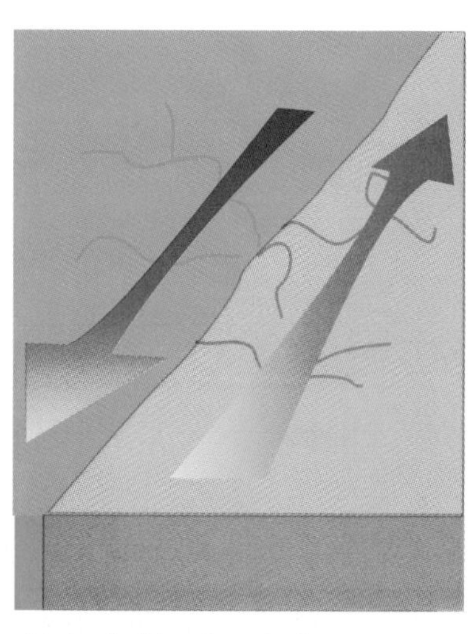

At last, the force becomes so great that the rocks suddenly break loose and slide past each other, causing an earthquake.

Most earthquakes happen along three fault areas: the Pacific belt, which includes California's San Andreas fault; the Alpide belt, which roughly follows the boundary between Europe and Asia; and the mid-ocean ridges, which lie deep in the Atlantic and Indian oceans. Most earthquakes—eighty percent—as well as ***volcanic eruptions*** occur in the Ring of Fire, a narrow band of land on the edges of the Pacific Plate.

About 900 earthquakes occur every hour. Most of these are very minor and probably can't be felt by humans.

ON A SCALE OF 1 TO 10: The Richter Scale

Some earthquakes go unnoticed. Others cause houses to shake, and a book or two, a vase, or a figurine to fall off the bookshelves. Still others destroy whole cities and landscapes.

Seismologists, the scientists who study earthquakes, use a scale to describe their force, or ***intensity.*** This scale is know as the ***Richter scale***. The scale was devised by Charles Richter and Beno Gutenberg. It measures earthquakes on a scale of 1 to 10, with 1 being the least powerful and 10 the most powerful.

A pendulum-type instrument called a ***seismometer*** measures the intensity of earthquakes. The base of the instrument is fixed firmly to the ground. A shaft or bar rises from the base. A weight hangs from the top of the shaft.

When shock waves move through the earth, the weight swings, leaving marks on the paper. The instrument is so sensitive, it can pick up movements from earthquakes thousands of miles away.

RICHTER NUMBER	DESCRIPTION OF EARTHQUAKE	NUMBER PER YEAR
2.0-3.4	mostly unnoticeable	800,000
3.5-4.2	barely felt	30,000
4.3-4.8	felt by most humans	4,800
4.9-5.4	felt by all	1,400
5.5-6.1	damaging	500
6.2-6.9	serious damage	100
7.0-7.3	very serious damage	15
7.4-8.0	catastrophic	4
8.0+	near total damage	1

MAJOR EARTHQUAKES

Date	Location	Deaths	Richter Measure
1906	San Francisco, U.S.	503	8.3
1906	Valparaiso, Chile	20,000	8.6
1920	Gansu, China	100,000	8.6
1923	Yokohoma, Japan	200,000	8.3
1927	Nan-Shan, China	200,000	8.3
1934	Bihar-Nepal, India	10,700	8.4
1939	Chillan, Chile	28,000	8.3
1946	Honshu, Japan	2,000	8.4
1960	Southern Chile	5,000	8.3
1964	Alaska, U.S.	131	8.4
1976	Tangshan, China	242,000	8.2
1977	Indonesia	200	8.0
1977	Northwest Argentina	100	8.2
1979	Indonesia	100	8.1
1985	Mexico City	4,200+	8.1
1989	San Francisco	62	6.9
1990	Northwest Iran	40,000	7.7
1990	Luzon, Philippines	1,621	7.7
1994	Los Angeles, U.S.	75+	5.8

Volcanic Eruptions

Before a volcano erupts, it "grows." Before we notice the eruption, the ***magma***, or molten rock, boils fifty to 100 miles below the surface. Then it begins to rise slowly through cracks and weak spots in the earth's surface. The ground under the magma grows from the force of the boiling rocks.

Most volcanoes are located near the boundaries of ***tectonic plates*** (see p. 47). Through these natural cracks magma finds its easiest passage to the earth's surface.

As magma rises, it gives off gases. These gases cause explosions that blow holes called ***vents*** or ***craters*** in the top of the volcano. Then ***lava,*** which is what magma is called once it reaches the earth's surface, rises up through the vents and flows down the mountainside like a fiery river. Lava reaches temperatures over 2,000° F (1,093° C.). In addition to spewing lava, volcanoes can also throw bits of rock and ash into the air. The larger rocks are called ***volcanic bombs.***

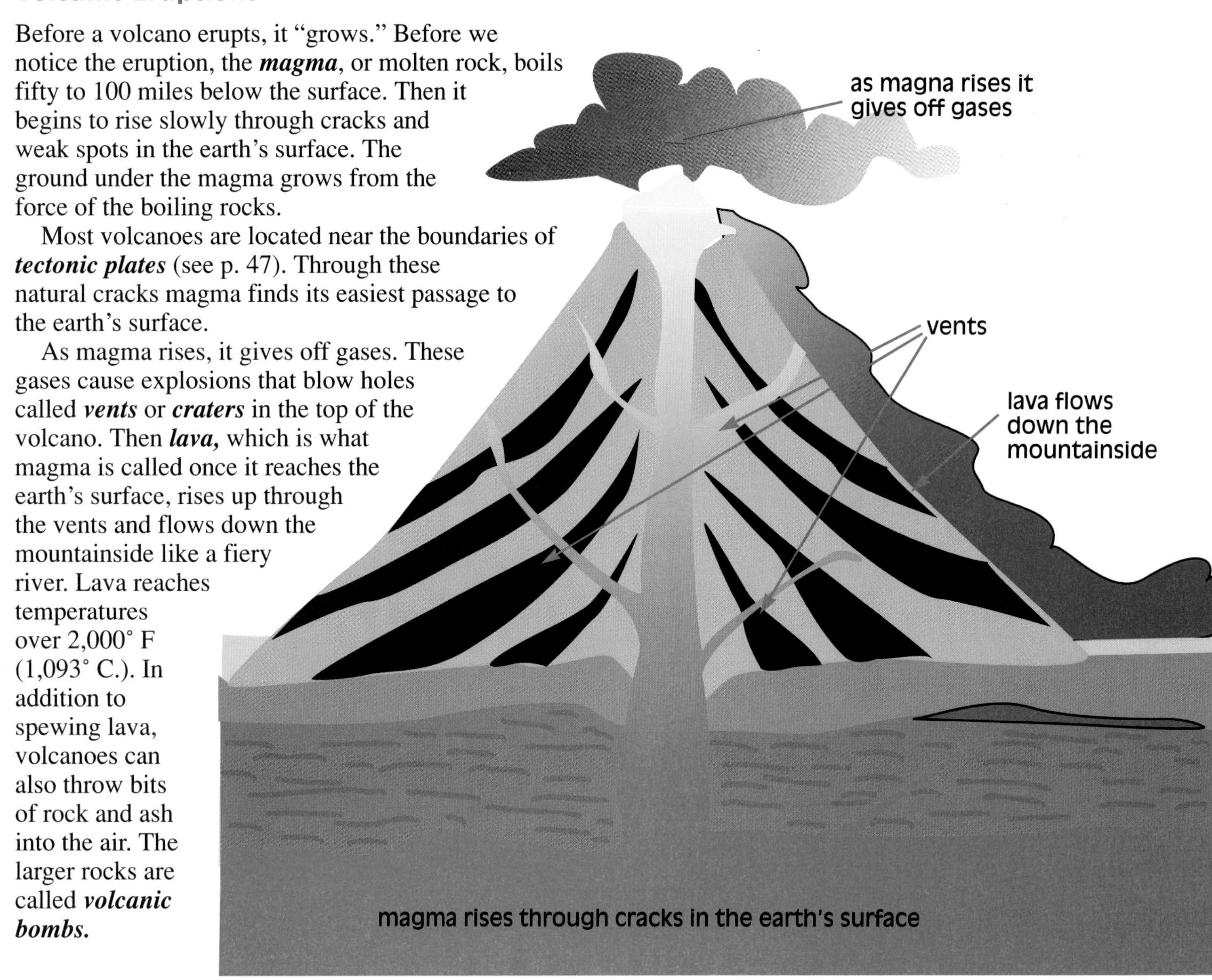

Not all volcanoes are in danger of erupting. In fact, some volcanoes haven't erupted in centuries. Scientists describe four types of volcanoes.

Four Types of Volcanoes

Active	Erupting constantly
Intermittent	Erupting at regular intervals
Dormant	Inactive, but may become active again
Extinct	Completely inactive for hundreds of years

Catch the Wave! Tsunamis, Earthquakes, and Volcanoes

The shock of an underwater earthquake or volcanic eruption affects the oceans as well as the land. ***Tsunamis*** are giant walls of water that rise from the ocean floor in response to tectonic shock. Tsunamis aren't noticeable at sea, but when they near shores, they can rise anywhere from six to sixty feet high.

Tsunamis travel at speeds of up to 500 to 600 miles per hour, and occur most often in the Pacific Ocean, affecting Japan, Hawaii, and the Aleutian Islands off Alaska.

Tsunamis are often confused with tidal waves. Tidal waves are usually high sea waves that can be caused by tsunamis, or by strong winds and storms.

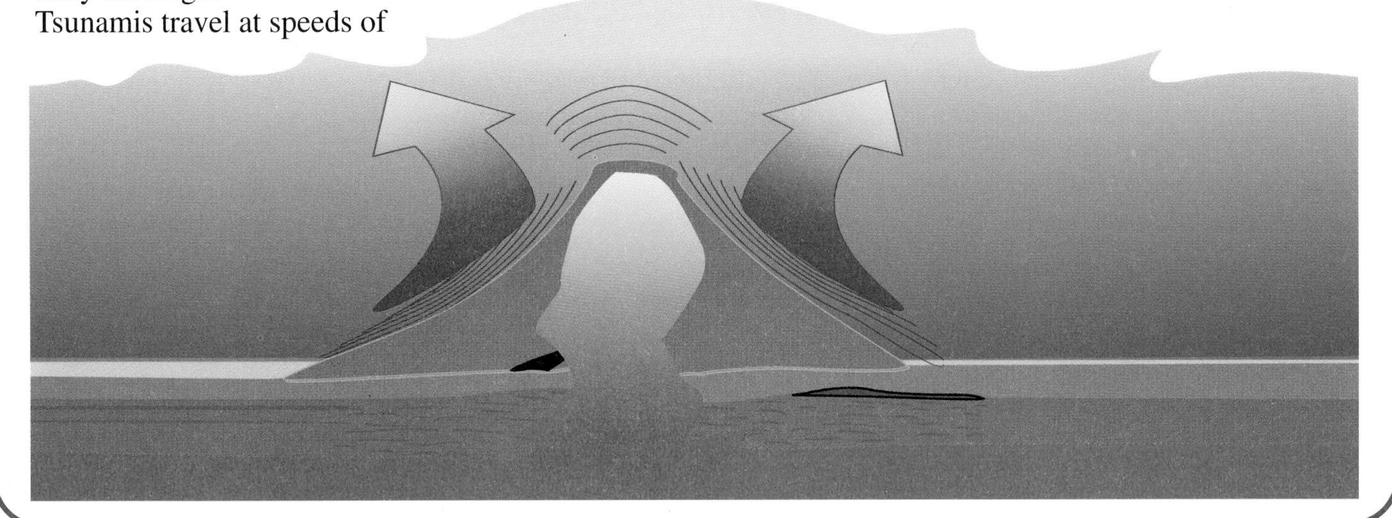

MAJOR VOLCANIC ERUPTIONS

Date	Volcano	Deaths	Date	Volcano	Deaths
79 A.D.	Mt. Vesuvius, Italy	16,000	1911	Mt.Taal, Philippines	1,400
1169	Mt. Etna, Sicily	15,000	1919	Mt. Kelud, Java	5,000
1631	Mt. Vesuvius, Italy	4,000	1951	Mt. Lamington, New Guinea	3,000
1669	Mt. Etna, Sicily	20,000	1966	Mt. Kelud, Java	1,000
1772	Mt. Papandayan, Java	3,000	1980	Mt. St. Helens, U.S.	60
1792	Mt. Unzen-Dake, Japan	12,000	1985	Nevado del Ruiz, Colombia	22,940
1815	Tamboro, Java	12,000	1986	Northwest Cameroon	app. 2,000
1883	Krakatau, Indonesia	35,000			
1902	Santa Maria, Guatemala	1,000			
1902	Mt. Pelee, Martinique	30,000			

3 The Water

Oceans and Seas

Almost three-fourths of the earth's surface is covered by ***oceans*** and ***seas***. An ocean is a large body of salt water. A sea is smaller than an ocean and is usually a part of an ocean.

Most of the earth's oceans and seas are connected to each other. Currents of water pass from one body of water to the next, both on the surface and in the dark, cold depths. The average ocean depth is just over two miles, although ocean trenches can be as deep as six miles.

Together, the oceans and seas contain nearly 317 million cubic miles of water. That's ninety-seven percent of all the world's water, including water vapor in the air, underground water, and ice caps and glaciers, as well as rivers, lakes, and other fresh water.

A **gulf** *is a large body of sa water partly surrounded by land. A* **bay** *is usually a small gulf, but names can' always be trusted. For example, the Gulf of California is much smaller than Hudson Bay.*

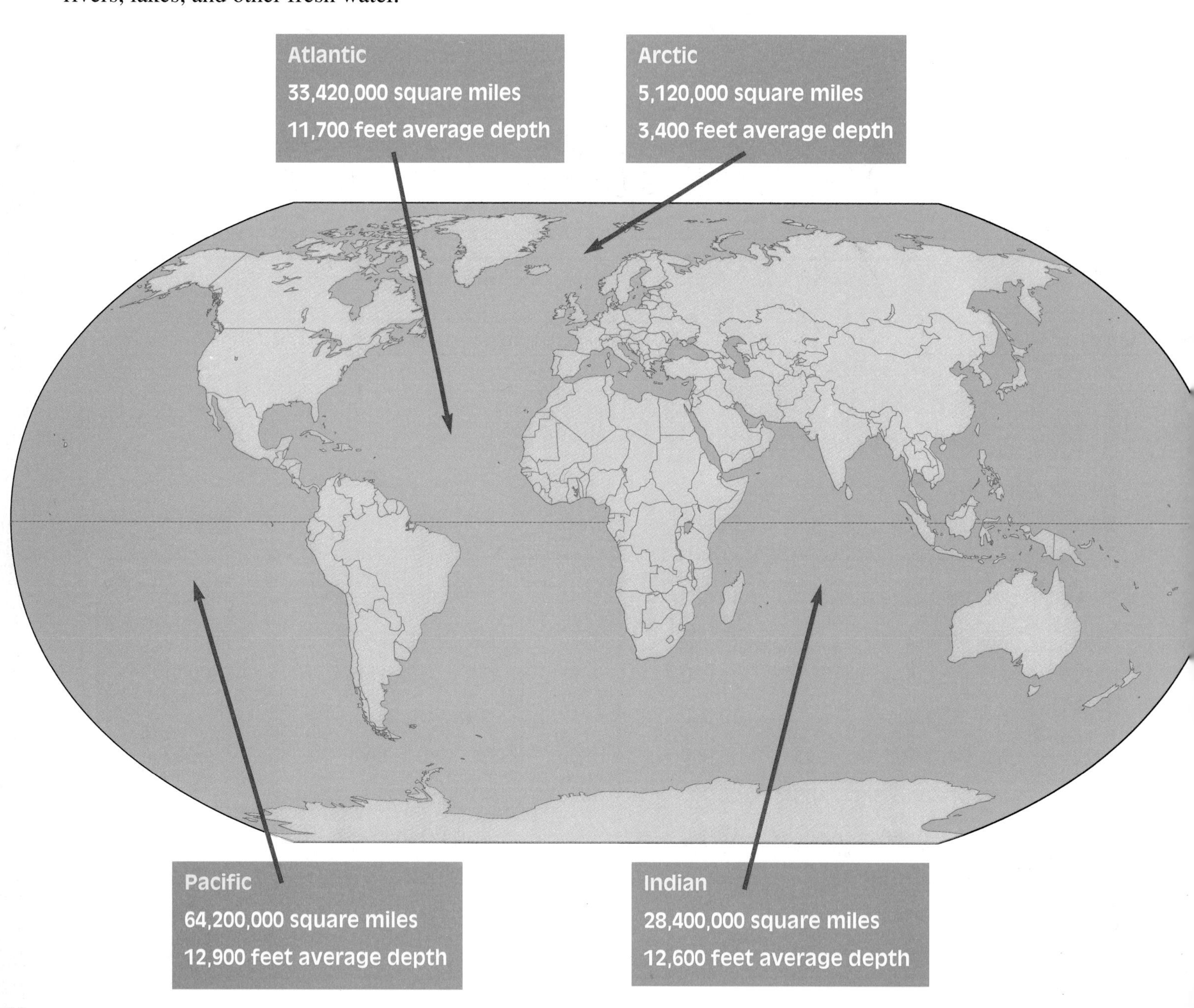

A Whole New World: The Ocean Floor

The ocean floor you see at the beach is usually a flat bed of sand. But the larger underwater landscape is filled with mountains, trenches, plateaus, and ridges, as well as vast, sandy plains.

Some underwater mountains rise thousands of feet from the sea floor, cresting above the water to form islands. Many of these mountain ranges contain ***volcanoes***, some active and some extinct (see p. 58).

Huge trenches, deeper than the Grand Canyon, fall between the mountains.

The ***continental shelf*** is a shallow, rocky "shelf" that slopes down from the shoreline before dropping off at a continent's edge. There the shelf drops off deeply into trenches or onto flat, sandy expanses called ***abyssal plains***.

Ridges rise gently in mid-ocean.

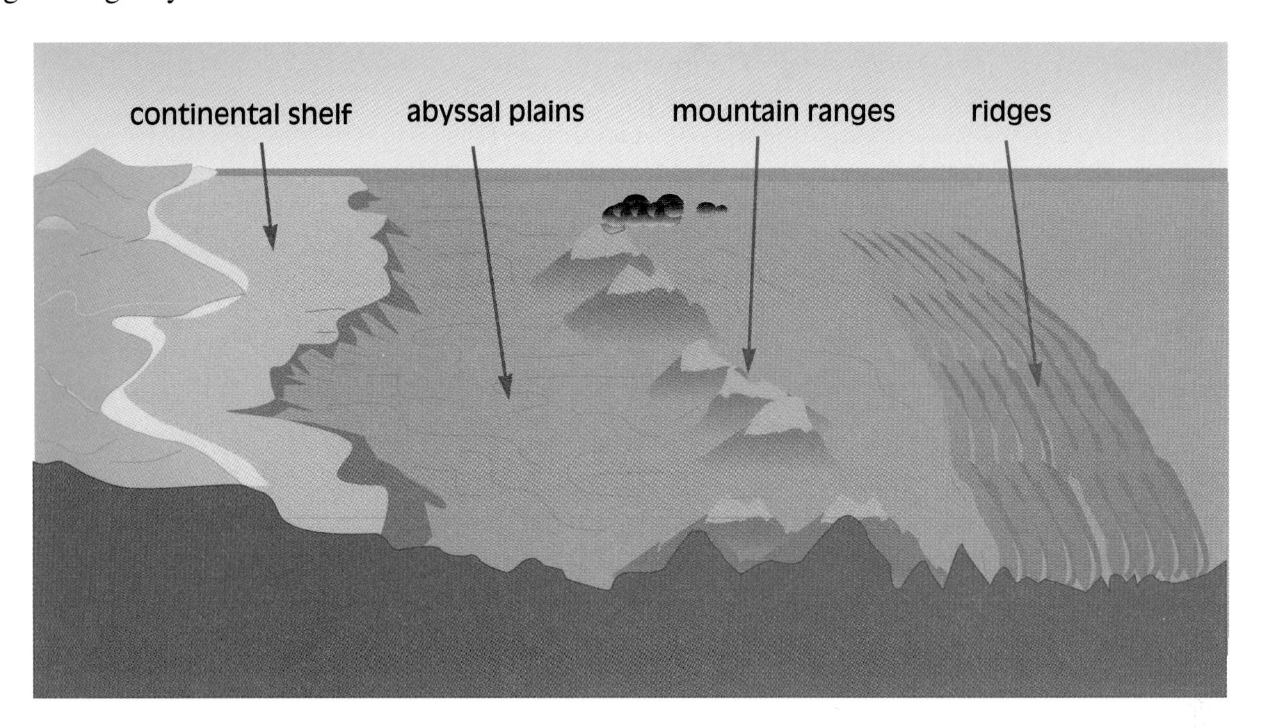

THE BRINY DEEP

	Depth (Feet)	Depth (Meters)
Pacific Ocean		
Mariana Trench	35,800	10,900
Tonga Trench	35,400	10,800
Philippine Trench	33,000	10,000
Kermadec Trench	33,000	10,000
Atlantic Ocean		
Puerto Rico Trench	28,200	8,600
South Sandwich Trench	27,313	8,300
Cayman Trench	24,700	7,500
Romanche Gap	21,700	6,600
Indian Ocean		
Java Trench	23,400	7,100
Ob' Trench	22,600	6,900
Diamantina Trench	21,700	6,600
Vema Trench	21,000	6,400
Arctic Ocean		
Eurasia Basin	17,900	5,500
Mediterranean Sea		
Ionian Basin	16,900	5,200

Ocean Zones

The oceans are divided from top to bottom into three layers, or ***zones***—the light, or ***photic***, zone, the twilight, or ***bathyl,*** zone, and the ***midnight*** zone, known as the abyss.

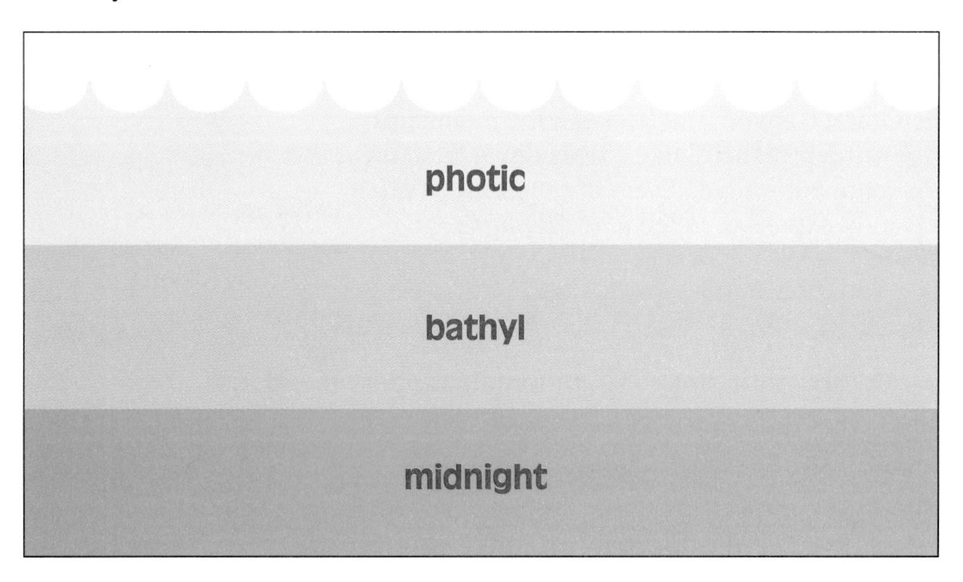

Ocean Currents

Surface and ***deep-water currents*** are flowing streams of water that run through the oceans, moving masses of warm and cold water to different regions.

Surface currents are caused by winds blowing across the water's surface. Deep-water currents are caused by the movement of dense, icy polar water toward the tropics and along the ocean bottom.

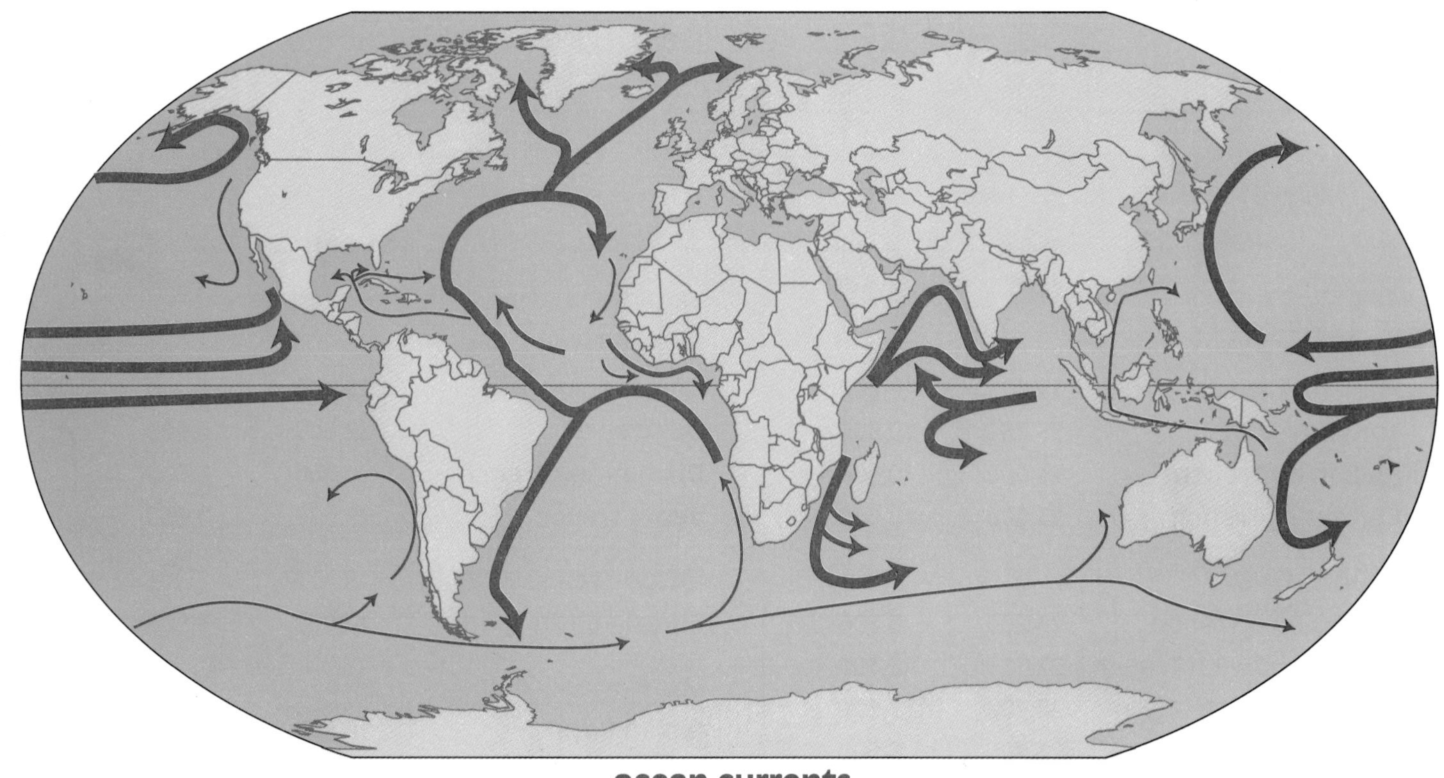

ocean currents

MAKING WAVES

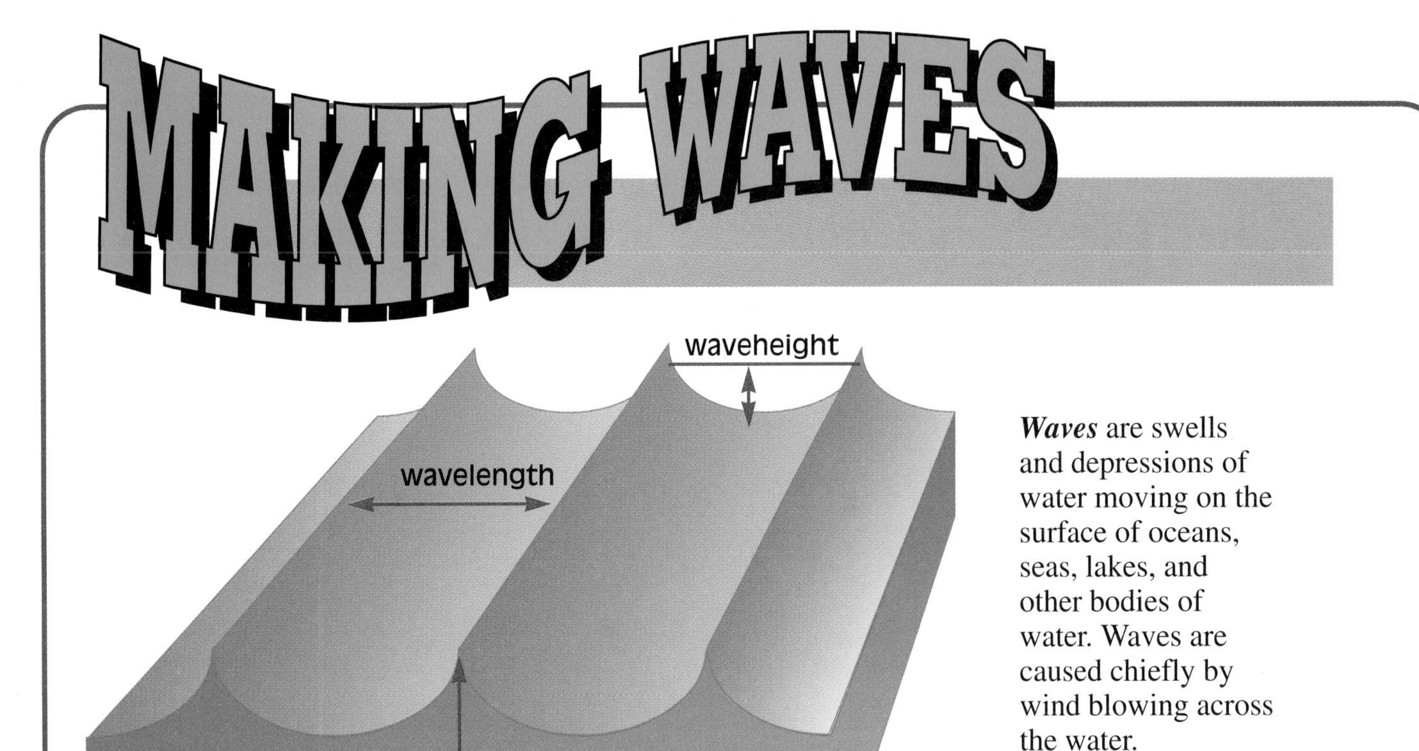

Waves are swells and depressions of water moving on the surface of oceans, seas, lakes, and other bodies of water. Waves are caused chiefly by wind blowing across the water.

Waves range in size from tiny ripples on a lake to giant ocean waves that reach over 100 feet high. The size of a wave depends on:

1. **The strength of the wind. The stronger the wind, the bigger the wave.**
2. **The duration of the wind. The longer the wind blows, the bigger the wave.**
3. **The fetch of the wind. The fetch is the distance of open water the wind blows over. The greater the fetch, the bigger the wave.**

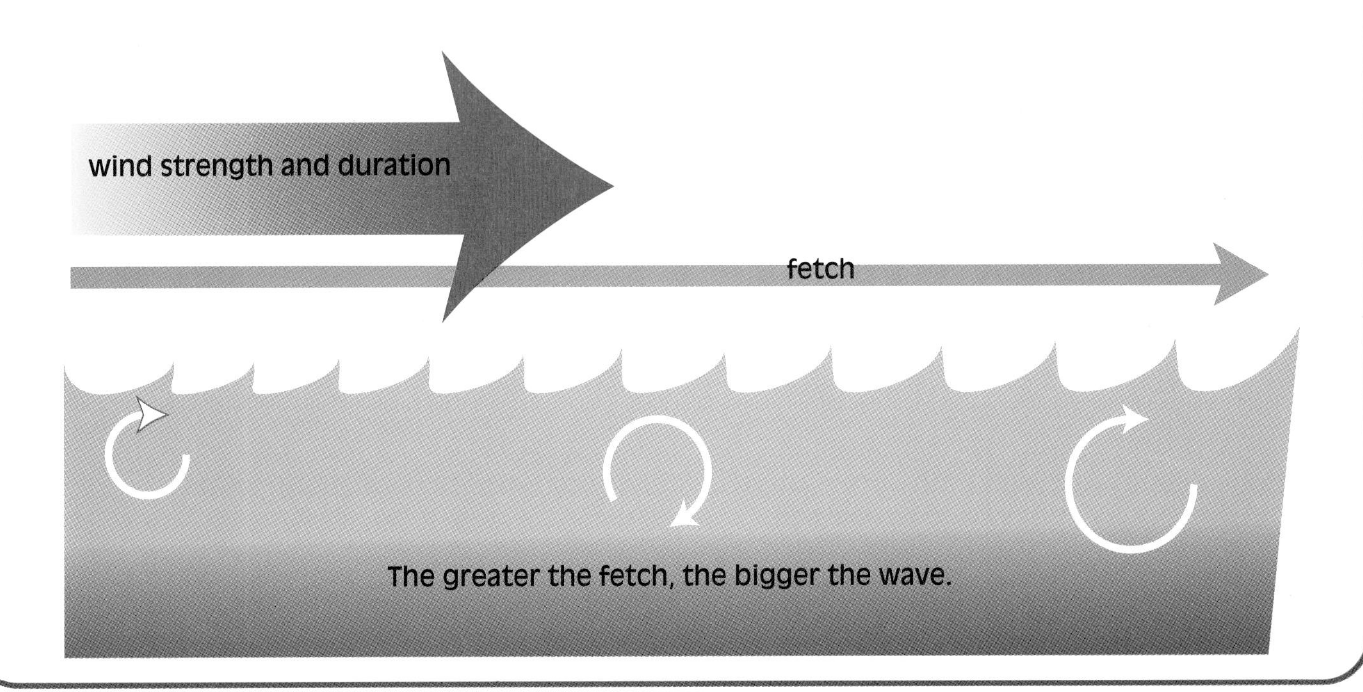

Tides

Tides are the regular rise and fall of waters in oceans and seas. They are caused by the pull of the moon's gravity on the earth. So as the moon moves in relation to the earth, the waters of the earth move, too. Tides rise and fall twice every lunar day, or every twenty-four hours and fifty-one minutes, the time it takes the moon to orbit the earth (see pp. 90-91). As the earth turns, the part of the oceans facing the moon will be at high tide.

Tidal Times

Low or ebb tide	Water is at its lowest point along the shore.
High or flood tide	Water is at its highest point along the shore.
High spring tide	Occurs twice a month during the new and full moons because the sun and moon are lined up and both "pull" on the earth.
Neap tide	Occurs during "half" moons because the moon and the sun are at 90° angles so the sun lessens the "pull" of the moon.

Lakes

Lakes are bodies of water surrounded by land. Lakes usually form in low spots on the land. Water from flooding, melting glacial ice deposits, rivers, and ground sources travels downhill to fill these low spots. Most lakes hold fresh water.

A Jump in the Lake

- Lake Superior, bordered by Minnesota, Wisconsin, Michigan, and Canada, is the world's largest freshwater lake, covering about 32,000 square miles.
- Lake Baikal, in Siberia, is the deepest lake in the world. It averages about 2,400 feet deep.
- Lake Titicaca, in the Andes Mountains in South America, is the highest lake that humans can travel on. It is approximately 12,000 feet above sea level.
- Lake Eyre, in Australia, covers almost 3,500 square miles, but disappears completely in years of low rainfall.
- Great Salt Lake in Utah is constantly evaporating. It once was three times the size it is today.

Ways in Which Lakes Are Formed

Ponds are small lakes.

Crater Lake

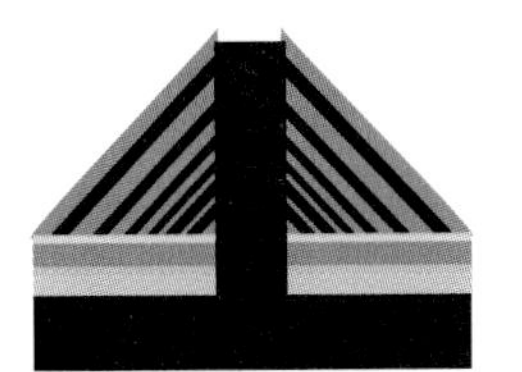

Water collects in craters left by volcanoes.

Glacial Lake

Ice from glaciers carves depressions (low areas) in the landscape. The ice melts and forms a lake.

Rift Valley Lake

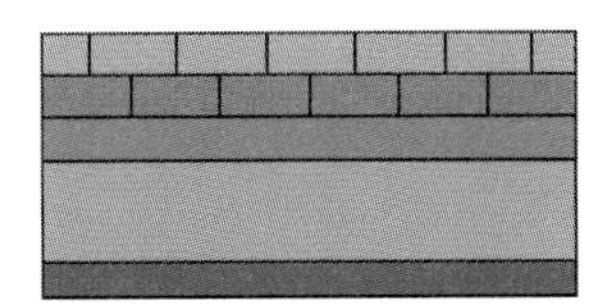

Shifts in plates on the earth's surface form depressions that fill with water.

Artificial Lake

People dig depressions and fill them with water — often diverted from rivers — to form lakes.

Rivers

Rivers are bodies of water that begin at a ***source*** and flow between ***banks*** of earth to a ***mouth***, where they empty into a larger body of water. Most large rivers have three parts, or ***courses***: the ***upper,*** the ***middle***, and the ***estuary.***

Many special river features can be seen as rivers flow through their courses.

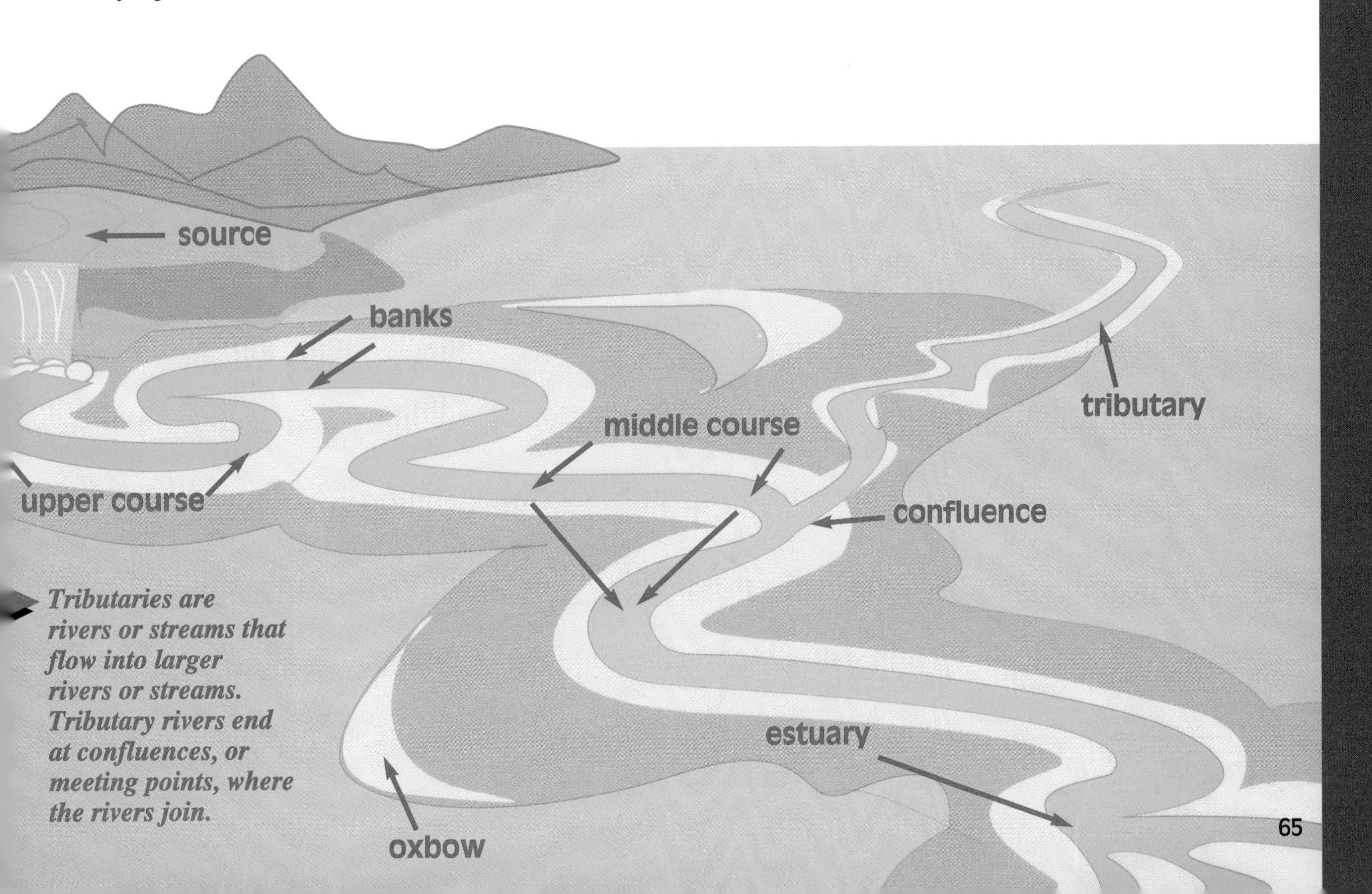

Tributaries are rivers or streams that flow into larger rivers or streams. Tributary rivers end at confluences, or meeting points, where the rivers join.

How Rivers Are Formed

Rivers form high above sea level in mountains or hills and drain rain and melted snow into larger bodies of water, such as bays and oceans. Water feeds into rivers in several ways: from underground springs, mountain streams, glacial run-offs, tributaries, and so on.

The World's Longest Rivers

RIVER	LENGTH IN MILES
North America	
Arkansas	1,460
Colorado	1,450
Columbia	1,243
Mackenzie	2,640
Mississippi	2,340
Missouri	2,320
Ohio-Allegheny	1,310
Pecos	926
Red	1,290
Rio Grande	1,900
Saskatchewan	1,200
St. Lawrence	800
South Platte	1,040
Yukon	1,979
South America	
Amazon	4,000
Japura	1,750
Madeira	2,010
Paran	2,400
Purus	2,100
Sao Francisco	1,990

RIVER	LENGTH IN MILES
Eurasia	
Amur	2,740
Chiang Jiang	3,960
Danube	1,780
Dnieper	1,420
Euphrates	1,700
Ganges	1,560
Huang (Yellow)	2,900
Indus	1,800
Irrawaddy	1,340
Lena	2,730
Mekong	2,600
Ob-Irtysh	3,360
Rhino	820
Tigris	1,180
Urul	1,580
Volga	2,194
Xi	1,200
Africa	
Congo	2,720
Niger	2,590
Nile	4,160
Zambezi	1,700

Streams, brooks, *and* **creeks** *are small rivers.*

Glaciers

Glaciers are slow-moving sheets of ice found in high mountain valleys and polar regions. Glaciers cover about six million square miles, or three percent, of the earth's surface.

Two Kinds of Glaciers

Valley glaciers fill high mountain valleys. They move downhill like giant rivers of ice, at a rate of about one inch each day.

Continental glaciers form on ice caps in freezing polar regions. They move during summer, when warmer temperatures melt the edges of the ice sheets and cause them to slide toward the sea. When continental glaciers meet ocean waters, huge blocks of ice break off, creating ***icebergs.***

How Glaciers Are Formed

Glaciers form where snow falls but doesn't melt because temperatures remain below freezing. More snow falls and builds up on the snow fields. Pressure from the weight of the snow finally turns the snow into huge sheets of ice, or glaciers.

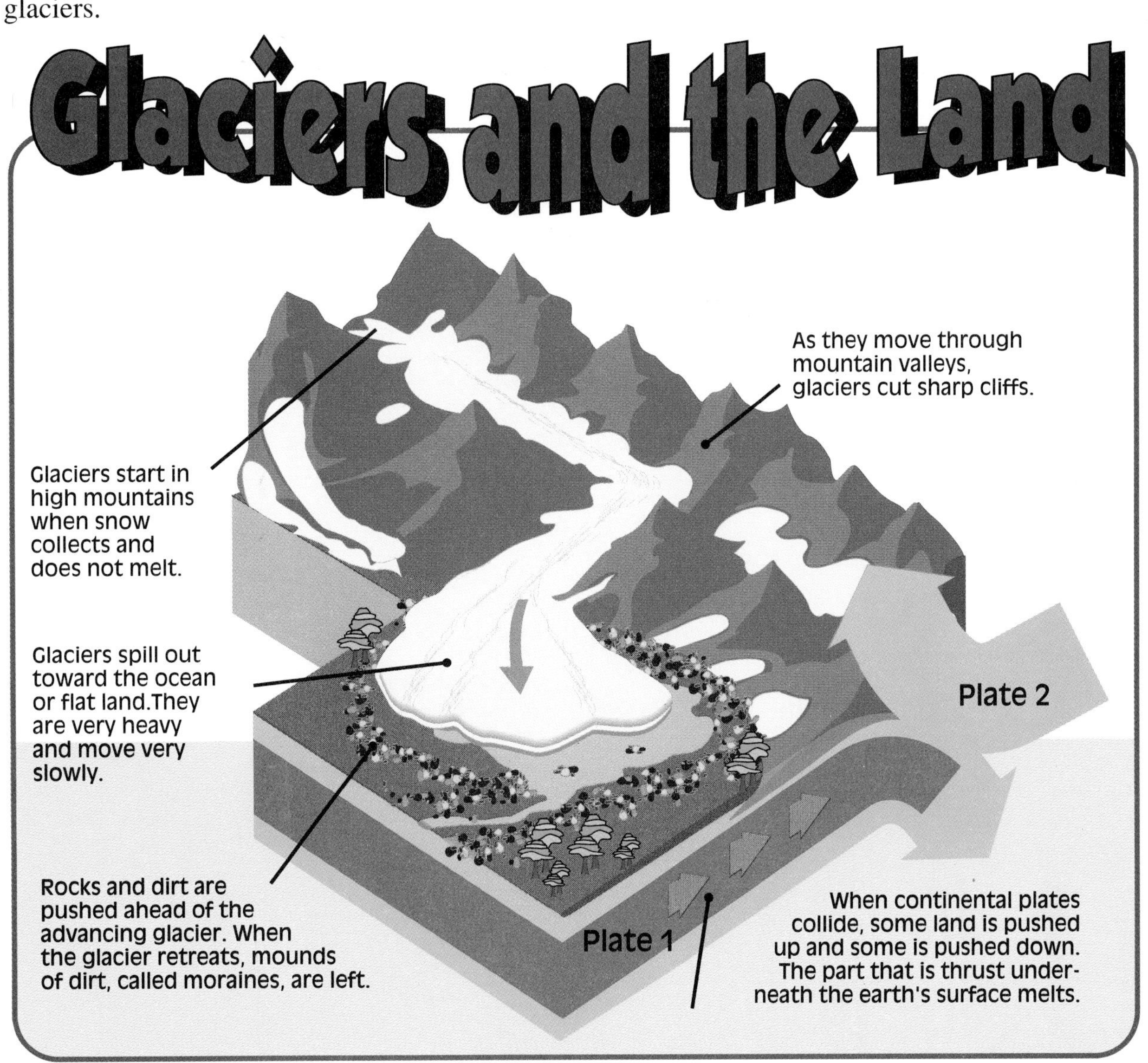
Glaciers and the Land
As they move through mountain valleys, glaciers cut sharp cliffs.
Glaciers start in high mountains when snow collects and does not melt.
Glaciers spill out toward the ocean or flat land.They are very heavy and move very slowly.
Plate 2
Rocks and dirt are pushed ahead of the advancing glacier. When the glacier retreats, mounds of dirt, called moraines, are left.
Plate 1
When continental plates collide, some land is pushed up and some is pushed down. The part that is thrust underneath the earth's surface melts.

4 The Atmosphere

The *atmosphere* is a layer of air that surrounds the earth's surface. The atmosphere is made up of nitrogen, oxygen, carbon dioxide, and water vapor, as well as tiny amounts of many other substances.

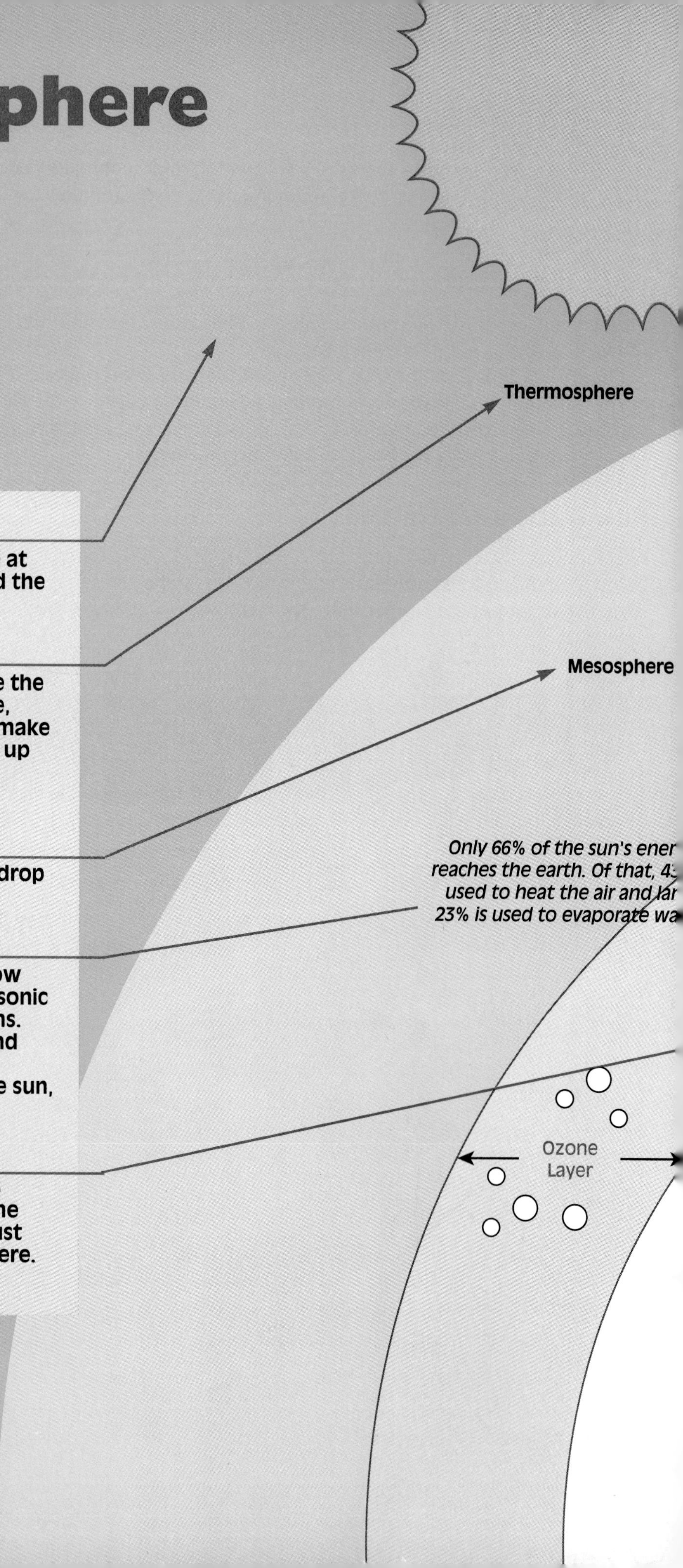

Thermosphere

Mesosphere

Only 66% of the sun's ener
reaches the earth. Of that, 4
used to heat the air and la
23% is used to evaporate wa

Ozone Layer

Exosphere

The border between the earth and space at about 310 miles. Satellites revolve around the earth in the exosphere.

Thermosphere

Ranges to about 150 miles or more above the earth's surface. Within the thermosphere, electrically charged particles called ions make up the ionosphere. Radio waves beamed up through the atmosphere bounce back to earth from the ion layers.

Mesosphere

Ranges to about 50 miles. Temperatures drop to under -100°F.

Stratosphere

Stretches to about 30 miles. Icy winds blow through the lower parts, speeding supersonic jets like the Concorde to their destinations. Above the clouds, the air is usually dry and clear. The ozone layer (see p. 42), which absorbs harmful ultraviolet rays from the sun, is here.

Troposphere

About 12 miles thick at the equator and 5 miles thick at the poles. More than half the atmosphere's gases, water vapor, and dust particles are in the first 4 miles. We live here. Clouds and weather form here, too.

Thirty-four percent of the sun's energy is turned back into space by the ozone layer, clouds, and dust in the air.

Stratosphere

Troposphere

Atmospheric Pressure

Atmospheric (or air) ***pressure*** is the weight of the air pressing down on the earth's surface. The miles of air above us weigh a lot. The air exerts about fifteen pounds of pressure on every square inch of your body.

Atmospheric pressure usually decreases as the amount of moisture in the air increases, because moist air is lighter than dry air.

Atmospheric pressure also decreases as ***altitude***, or height above the earth's surface, increases. The cabins of high-flying jets are "pressurized" in flight to keep the air pressure inside the plane the same as it is on the ground. Otherwise, it would be difficult—or even impossible—to breathe.

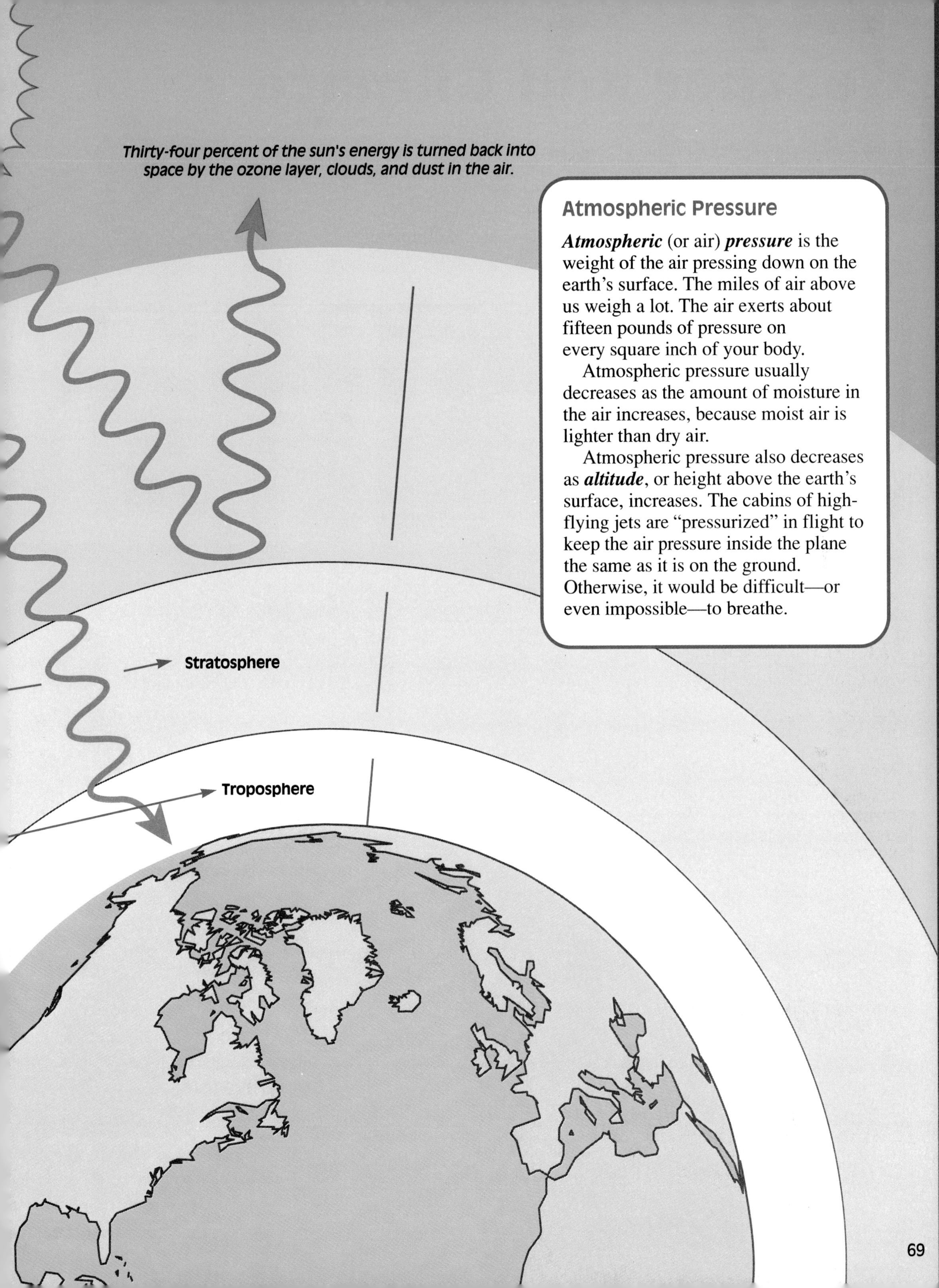

5 Weather and Climate

Weather is the day-to-day variation in temperature, winds, precipitation, and other atmospheric conditions. While weather can range from clear skies and heat waves to blizzards and thunderstorms, weather conditions are determined by just three basic ingredients: heat, air, and water.

Climate is the usual weather in an area over a long period of time.

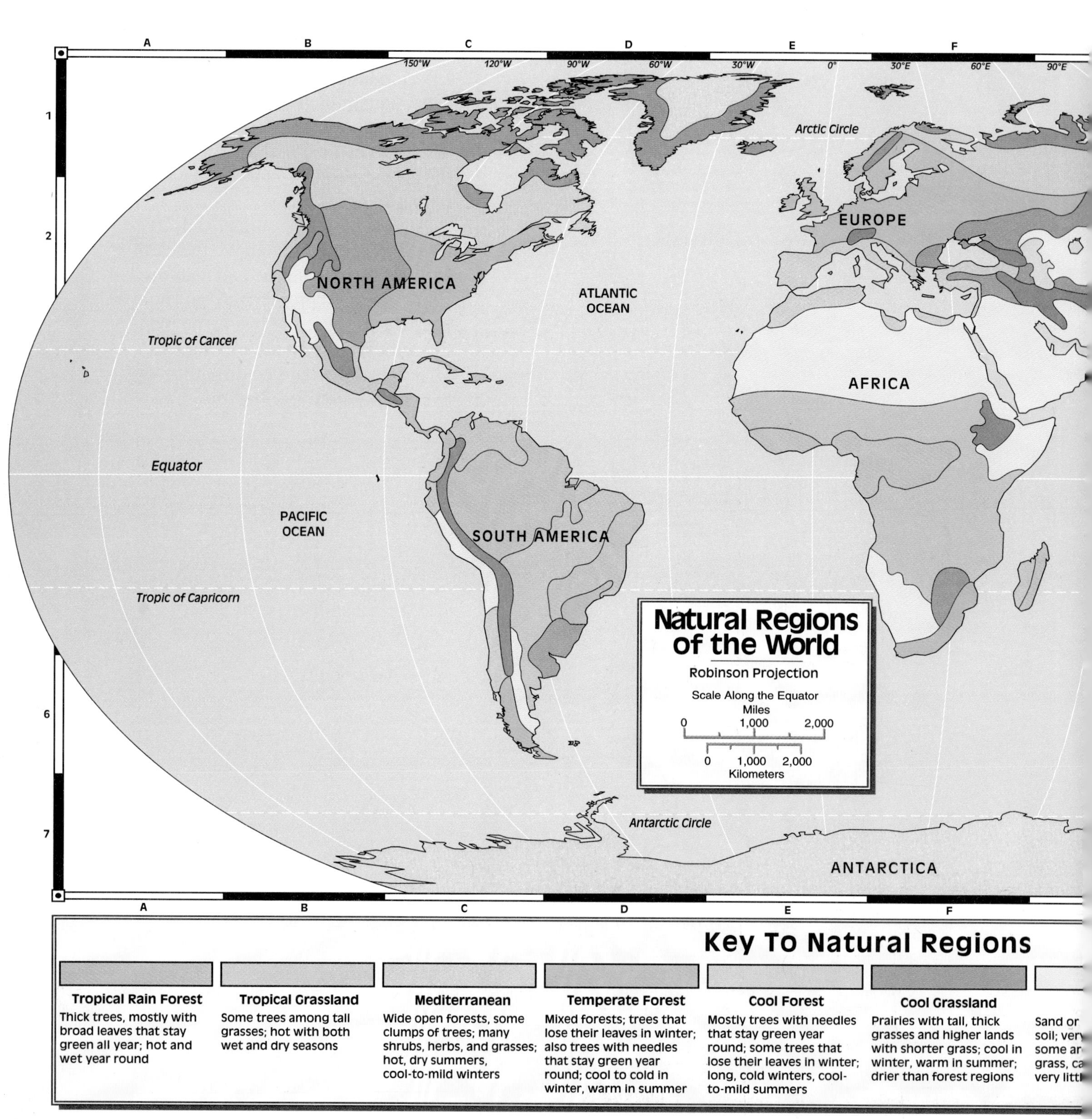

Weather Words: Talk Like the TV Forecaster

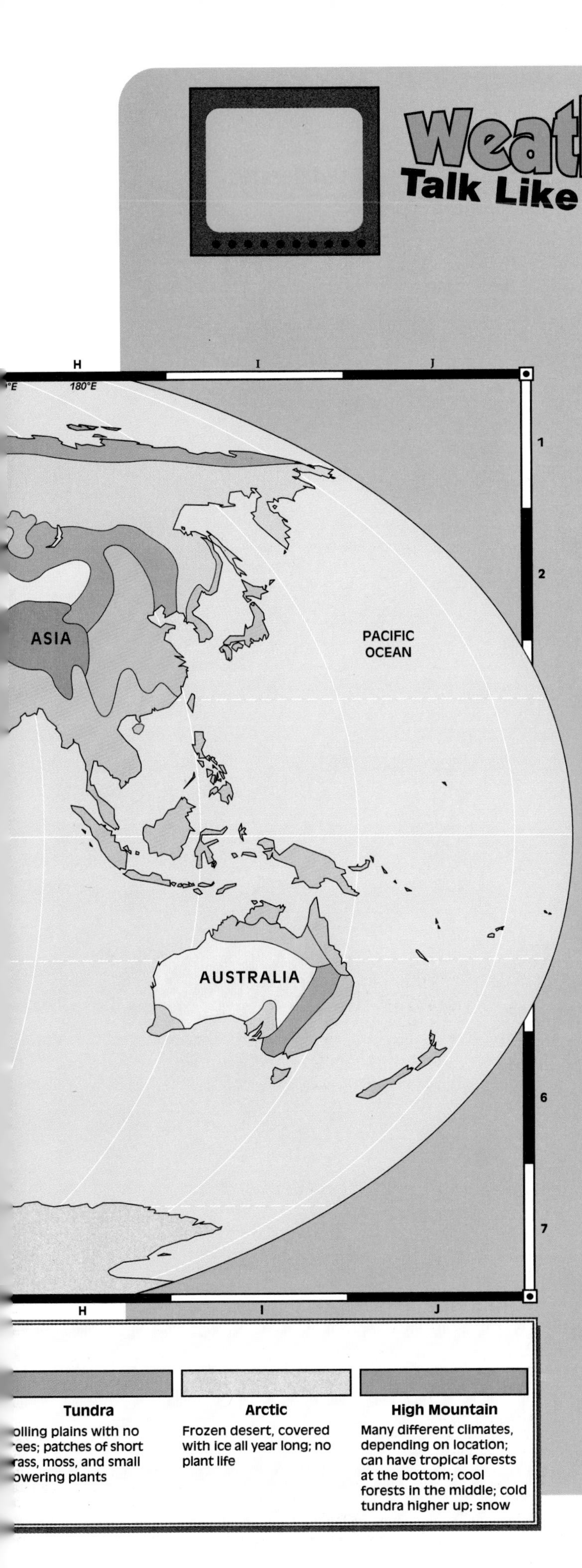
H
I
J
180°E
1
2
ASIA
PACIFIC
OCEAN
AUSTRALIA
6
7
H
I
J
Tundra
olling plains with no
ees; patches of short
rass, moss, and small
owering plants
Arctic
Frozen desert, covered with ice all year long; no plant life
High Mountain
Many different climates, depending on location; can have tropical forests at the bottom; cool forests in the middle; cold tundra higher up; snow

Air Mass
A huge body of air spread over hundreds of miles in which the temperature and humidity are nearly the same throughout, for example, "an area of low pressure," "a mass of warm air," and so on.

Air Temperature
The measure of hotness and coldness in the air.

Atmospheric Pressure
The weight of the atmosphere, or the pressure the atmosphere puts on the earth's surface. Changes in air pressure cause changes in the winds and affect cloud formation. Air pressure also provides clues in weather forecasting. (See also p. 69.)

Dew Point
The temperature at which water vapor changes into liquid water when it comes into contact with a surface, for example, lawn grass.

Fog
A mass of tiny water droplets in the air near the ground.

Front
The boundary between two air masses (see p. 73).

Humidity
The amount of moisture, or water vapor, in the air.

Precipitation
Any moisture that falls from the sky. Precipitation includes ***rain, mist, snow, sleet,*** and ***hail.*** (See also p.77.)

Wind
The movement of air across the atmosphere.

Heat from the Sun

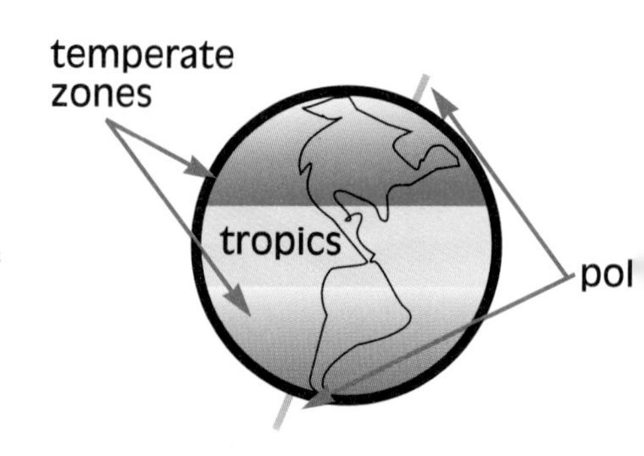

Sunlight heats the earth's surface, which warms the air above it. But the sun does not heat the earth evenly. The ***poles*** are frigid, and the ***tropics*** (the band surrounding the equator) are usually hot. That's because the tropics face more directly toward the sun than the poles.

In the ***temperate zones*** (the areas between the tropics and the poles), the amount of heat received from the sun varies greatly during the year because the earth tilts on its axis as it orbits the sun. The variation causes the seasons: summer, fall, winter, and spring. The seasons north of the equator are the opposite of the seasons south of the equator. When it's summer in the United States, it's winter in Brazil.

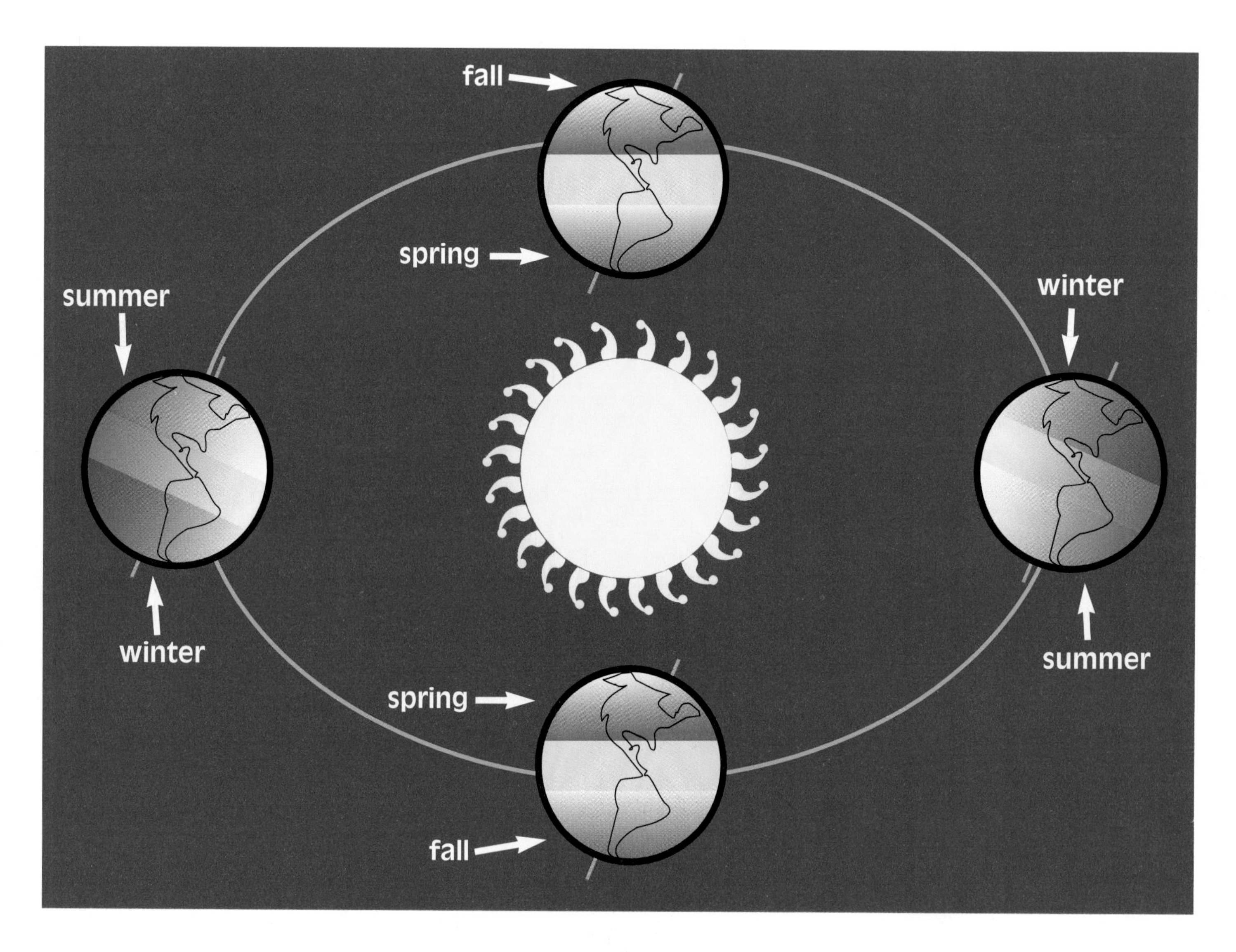

Air

Air Masses and Fronts

Sunlight heats the earth's surface in some places more than in others. The warm ground and water heat the surrounding air. This uneven heating creates ***air masses***, large bodies of air, at different temperatures and atmospheric pressures. The boundaries between air masses are called ***fronts***.

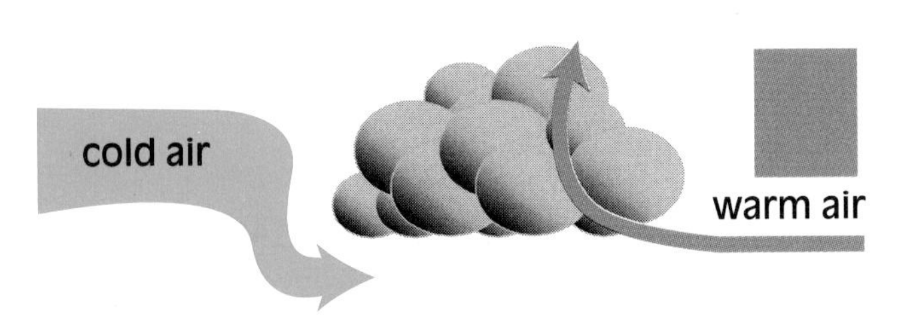

Cold Front

Cold air moves in on an area of warm air. The heavier cold air slides in underneath the lighter warm air mass and pushes it up. ***Clouds*** (see p. 76) and ***thunderstorms*** (see p. 78) often form.

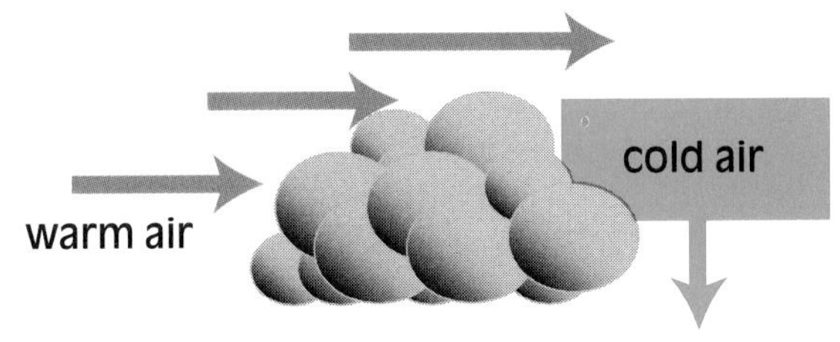

Warm Front

Warm air moves in on an area of cold air. The lighter warm air slides over the heavy, cold air, creating a front with a gentle slope. Clouds form, usually leading to some form of ***precipitation*** (see p. 77).

Stationary Front

Cold and warm air masses meet, but neither moves on the other. Clouds often form at the boundary.

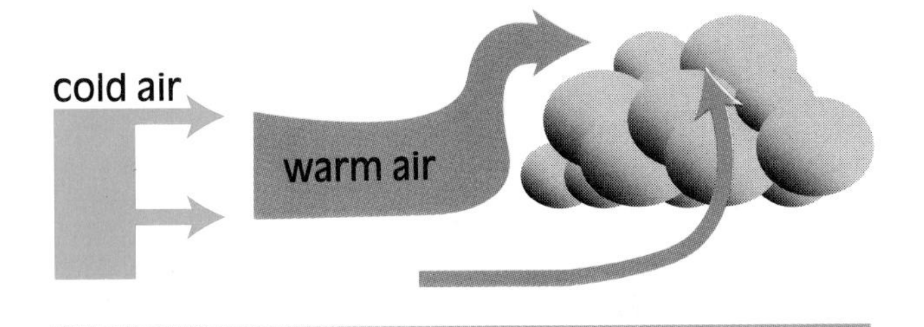

Occluded Fronts

When a cold front catches up to a warm front, the cold air forces the warm air up. Clouds and precipitation result.

Wind and Wind Patterns

Wind is moving air. It travels in huge spirals from areas of high atmospheric pressure to areas of low pressure. Wind can also move ***horizontally*** across the earth's surface near the poles and the equator. These different movements cause complicated wind patterns across the planet.

Wind ***velocity***, or speed, is estimated using the ***Beaufort scale***. The scale divides winds into categories, or ***forces***, based on wind strength.

Local winds occur in particular parts of the world. Here are some of their special names:

Chinook

Warm, dry winds that flow down the side of a mountain are called chinooks in the western United States, where they flow down the Rockies and the Cascades, and ***foehns*** in Europe, where they flow down the Alps.

Harmattan

These cool, very dry winds form over the Sahara in Northern Africa and blow west or southwest to the African coast from December through February bringing relief to the tropics.

Monsoon

These rainy winds that control the climate of Asia are the result of the heating of the land in the summer and its cooling off during the winter.

Sirocco

The name for these rains that blow north from the Sahara is Italian. Sirocco winds may pick up moisture over the Mediterranean Sea and arrive warm and damp in Europe, or they may stay dry and darken the sky with their sand.

BEAUFORT SCALE

Beaufort Number	Mph	Description
0	below 1	Calm
1	1-3	Light air movement
2	4-7	Light breeze
3	8-12	Gentle breeze
4	13-18	Moderate breeze
5	19-24	Brisk
6	25-31	Strong breeze
7	32-38	Moderate gale
8	39-46	Gale
9	47-54	Strong gale
10	55-63	Storm
11	64-73	Violent storm
12-17	74+	Hurricane

World's Fastest Winds: Jet Streams

In the *stratosphere* (see p. 68), winds blow at high speeds. In fact, if you took a ride on the jet stream, you'd travel at 150 miles per hour—or more!

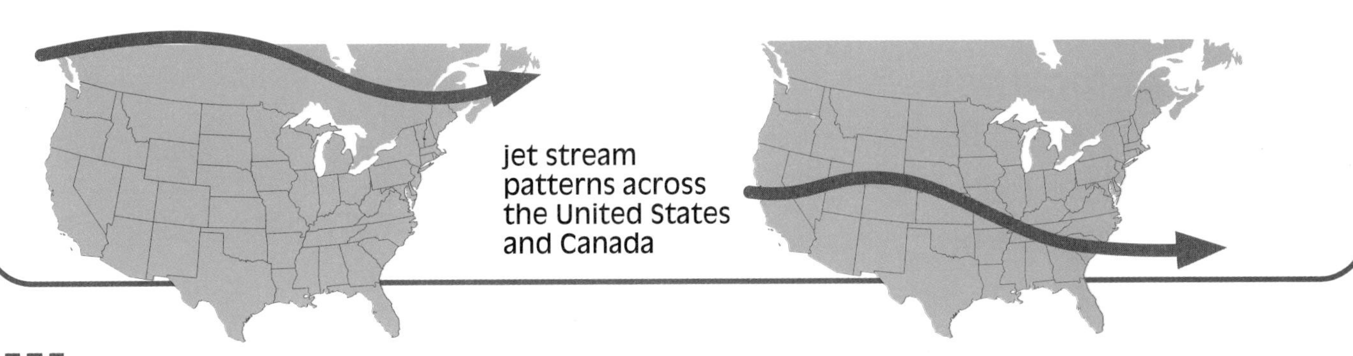

jet stream
patterns across
the United States
and Canada

Water

On any day, about forty trillion gallons of water—about ten times the amount of water in all the world's rivers—are in the atmosphere in the form of a gas called ***water vapor***. This gas is formed when liquid water ***evaporates***. The amount of water vapor in the air is called ***humidity***.

Water vapor condenses in the atmosphere in clouds or fog as droplets or ice crystals. When the droplets or ice crystals become too heavy to be held in the air, they fall back to the earth in the form of ***precipitation***.

Water evaporates,
condenses to form
clouds, and falls back
to the earth.

CLOUDS

Cirrus (20,000–40,000 feet)

Cirrostratus (20,000–40,000 feet)

Cirrocumulus (20,000–40,000 feet)

Altostratus (6,000–20,000 feet)

Altocumulus (6,000–20,000 feet)

Stratocumulus (below 6,000 feet)

Stratus (below 6,000 feet)

Cumulus (below 6,000 feet)

Cumulonimbus (cloud mass from below 6,000 feet to over 50,000 feet)

Cumulonimbus

Cirrus

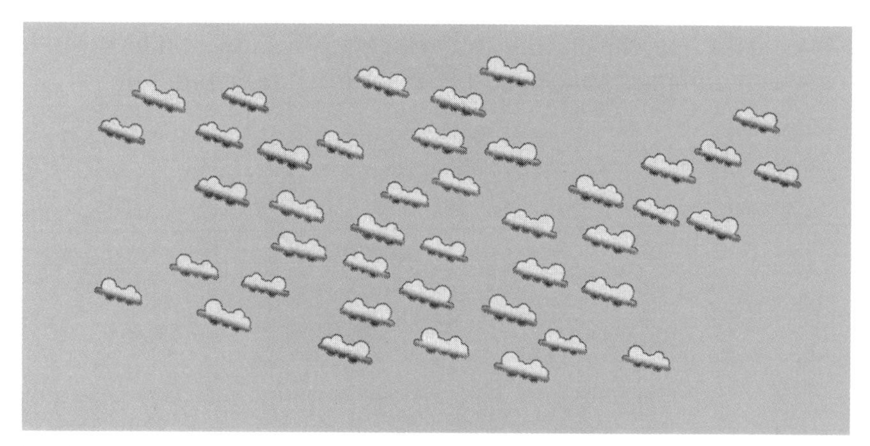

Stratus

Cumulus

Stratocumulus

Predicting the Weather

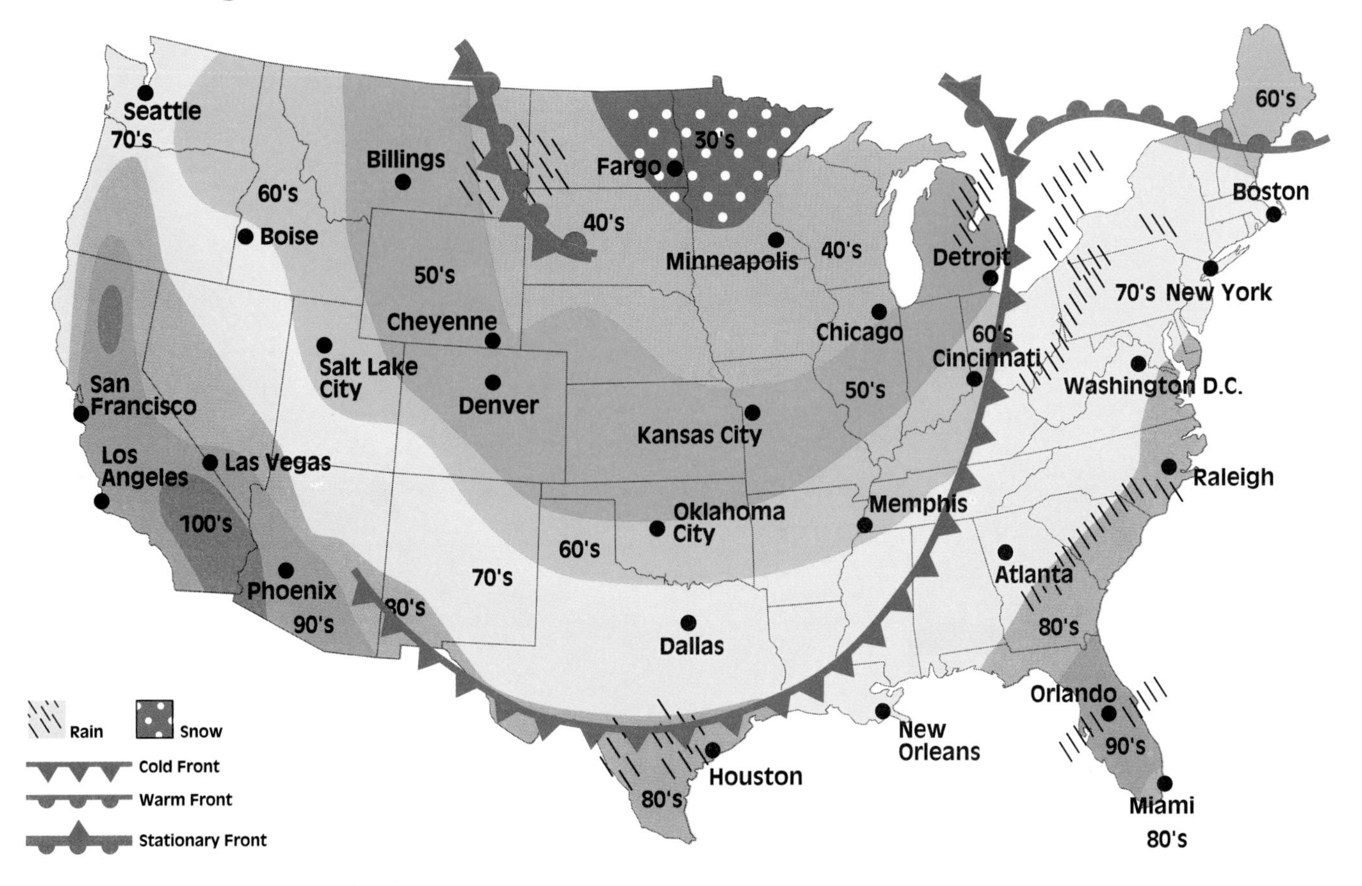

Dew
Water that condenses on surfaces when the temperature cools enough to change water vapor into liquid water—the ***dew point.***

Drizzle
A light rain of tiny water droplets. Drizzle is easier to feel than to see.

Fog
Clouds formed at ground level. Fog is not really a form of precipitation.

Freezing Rain
Rain that freezes when it lands, often coating buildings, trees, roads, and cars with a layer of ice.

Hail
Ice balls that fall during some thunderstorms. Some hailstones are tiny, others are larger than baseballs.

Rain
Water drops that fall steadily in temperatures above freezing.

Sleet
Raindrops that freeze into ice pellets before they hit the ground.

Snow
Ice crystals formed directly into snowflakes from condensing water vapor that falls when the temperature is below freezing.

(See also Storms, p. 78)

Storms

Rainstorms are heavy rains accompanied by high winds. ***Snowstorms*** occur when snow falls from clouds. ***Blizzards*** are snowstorms accompanied by high winds.

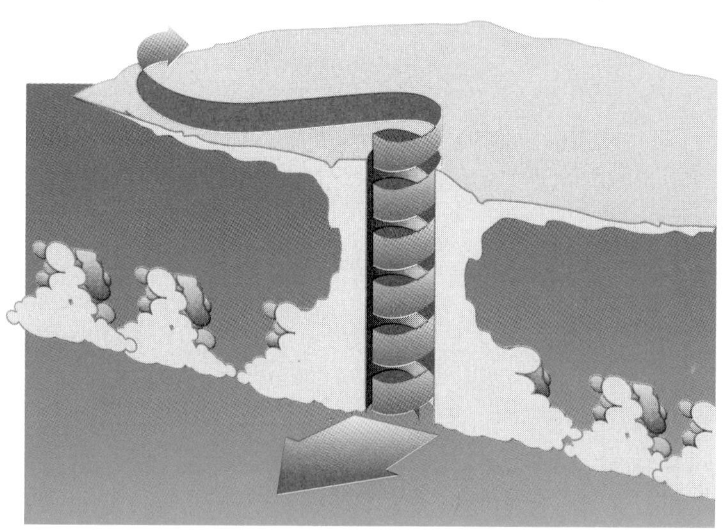

Cyclones are warm, low-pressure weather systems surrounded by cooler air. When the winds of a cyclone exceed seventy-five miles per hour, the cyclone is called a ***hurricane***. A ***typhoon*** is the name given to hurricanes in the region of the Pacific Ocean west of the International Date Line.

Thunderstorms occur when a warm, moist air mass near the ground is covered by a mass of cold air. When winds of fifty-eight miles per hour or greater blow, a thunderstorm becomes severe. The winds can create a strong, rotating column of air that reaches from a cumulonimbus cloud to the ground. This column is known as a ***tornado***.

Stars and Galaxies

The Big Bang Theory
How the Universe Was Made

The ***universe***, or all the stars, planets, and everything else in space, is a very mysterious place. No one really knows how big it is, or when it came to be.

Scientists have tried to explain how the universe was formed. Many agree that the Big Bang theory provides a good explanation. According to the theory, the universe was formed about fourteen billion years ago in one giant explosion—or "big bang." From the explosion, the bits of gas and matter that form the stars and planets flew out in all directions. Some scientists think that the explosion continues to this day, which explains why billions of stars, planets, and other heavenly bodies are constantly moving deeper and deeper into space.

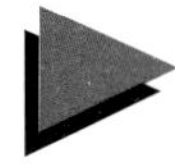

The study of outer space and its bodies is called **astronomy**. *The scientists who study space are* **astronomers**.

Stars

How Stars Are Formed

Stars are formed when gravity pulls together clouds of gas and dust in deep space. As more gas and dust get pulled in, the clump heats up. Finally, heat and pressure make a fire, called ***nuclear fusion*** (see Atoms, p. 97).

What Stars Are Made Of

Stars are made up of many gaseous elements (see Elements, pp. 98-99), including hydrogen, helium, nitrogen, oxygen, iron, nickel, and silicon. Hydrogen is their major fuel and causes the star to "shine," or give off light and heat energy.

Why Stars Are Bright

Some stars seem "brighter" than others. The brightness of a star is called its ***magnitude***, or size. Big stars are brighter than little stars, and so have greater ***absolute*** magnitude. Stars close to Earth also appear brighter than those farther away. They have greater ***apparent*** magnitude.

The Five Brightest Stars

Star	Magnitude	Distance from Earth in Light Years
Arcturus	-0.06	36
Canopus	-0.73	650
Rigel	-0.20	540
Sirius	-0.60	9
Vega	-0.04	27

Astronomers measure the brightness of stars based on their absolute magnitude. Brighter stars have smaller magnitude numbers. The brightest sky objects have negative magnitude numbers. So while our sun (see pp. 83-84) has an apparent magnitude of -27, it has an absolute magnitude of only +5.

Stars have different colors, depending on how hot they are. Very hot stars are blue, and the coolest stars are red. Usually, larger stars are brighter and hotter.

Surface Temperature	Color	Example
54,000°F (30,000°C)	Blue	Zeta Puppis
27,000-54,000°F (15,000-30,000°C)	Blue-white	Rigel
14,500-27,000°F (8,000-15,000°C)	White	Vega
About 13,500°F (7,500°C)	White-yellow	Canopus
About 11,000°F (6,000°C)	Yellow	Sun
About 9,000°F (5,000°C)	Orange	Aldebaran
About 6,000°F (3,000°C)	Red	Betelgeuse

Giants and Supergiants
The biggest and, usually, the brightest stars. These stars range from two- to three-thousand times larger in diameter than the sun, and hundreds to thousands of times brighter. The gases in giants and supergiants are much more spread out than those in smaller stars.

Dwarfs
The tiniest stars, less than one-hundredth the size of the sun. ***White dwarf*** stars are ***dying*** stars that are extremely dense.

Variable Stars
Stars that change in brightness. Some variable stars change in a predictable pattern, while ***irregular variables*** do not.

Star Clusters
Stars often form in groups called ***clusters.*** Some are made up of loosely grouped stars. Others, called ***globular clusters,*** are stars packed into a tight ball, or ***globe***.

Double Stars
Most stars form in pairs or larger groups. Gravity (see p. 88) keeps these stars in orbit around each other. Double stars are also called ***binary*** stars.

Novas and Supernovas
Exploding stars. When energy in a star builds up faster than the star can give it off in heat and light, the star explodes. A bigger than normal explosion creates a ***supernova***.

Black Holes
Objects whose force of gravity is so strong at the surface that no light waves can escape. Some black holes are collapsed stars, and others may be as small as subatomic particles (see p. 100).

Dying Stars
Stars that have cooled and been pulled together by gravity. ***White dwarfs, neutron stars,*** and ***pulsars*** are three types of dying stars.

Neutron Stars
Extremely dense dying stars. As a massive star changes from being a ***supernova***, its core consists mainly of neutrons (see p. 97). This core has an extremely heavy mass (see pp. 95-96).

Pulsars
Rapidly spinning ***neutron stars*** that give off light and radio waves. Pulsars are thought to be formed during the explosion of ***supernovas***.

Quasars
Short for ***quasi*** (almost)-***stellar radio source***. Quasars are mysterious starlike bodies, although scientists are still uncertain about what they really are.

Galaxies

A ***galaxy*** is a large group of ***stars***. We live in the ***Milky Way*** galaxy, a group of stars about 100,000 light-years across and about 1,300 light-years thick. The Milky Way's many "arms" that spread from a common center make it a ***spiral galaxy.*** Galaxies with oval shapes are called ***elliptical*** galaxies. Galaxies with no particular shape are called ***irregular*** galaxies.

Galaxies exist in groups. The Milky Way is part of the ***Local Group***. It is the second largest galaxy in the group, after the Andromeda Galaxy, a spiral galaxy that measures about 130,000 light-years across.

The Milky Way is a spiral galaxy. So is Andromeda. In the Local Group, there are three spiral galaxies. Most of the rest of the 30 galaxies in the Local Group are elliptical dwarfs.

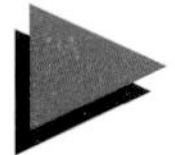

What's a light-year*? Distances in space are so great that they can't be measured in inches, feet, or even miles. Instead, scientists use light-years to measure these distances. A light-year equals the distance that light travels through space in one year—about six trillion miles!*

The Sun

Like other stars, the sun is a giant ball of hot gas. It gives off light and heat. About ninety-three million miles away, the sun is the closest star to Earth. The next closest star, Proxima Centauri, is about twenty-five trillion miles away.

The sun is about five billion years old. It burns off more than twenty-four billion tons of hydrogen gas each minute. In another five billion years, as the sun begins to die, it will expand and burn away life on our planet.

Age	5 billion years
Mass	332,000 times Earth's mass
Diameter	865,000 miles
Surface Temperature	6,000°F
Distance	93 million miles from Earth

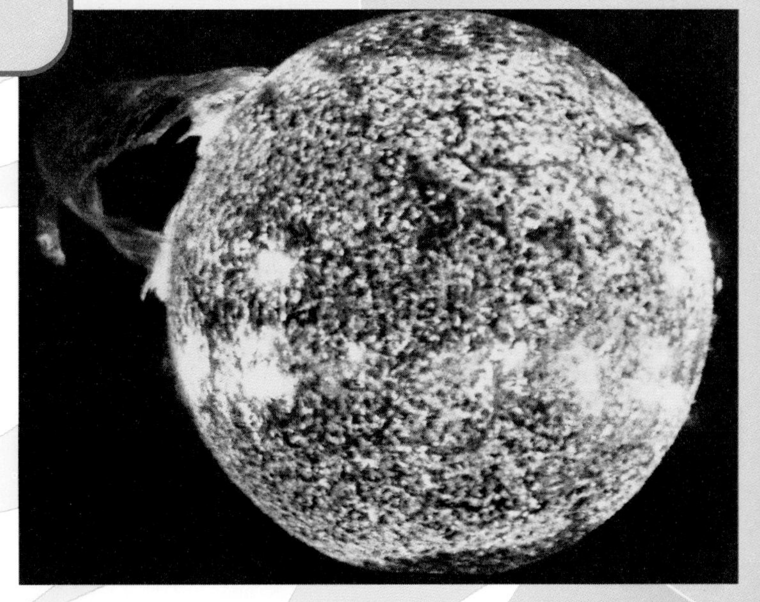

Never look into the sun, either directly or through binoculars or a telescope. Because the sun is so bright, it can seriously damage your vision, or even blind you.

Parts of the Sun

Because looking directly at the sun will burn your eyes, scientists have developed many tools for observing the sun indirectly. These tools have enabled scientists to learn a great deal about the sun's makeup and atmosphere.

SUNSPOTS

The surface of the sun flames brightly with hot gases. But sometimes the sun's magnetic field disturbs the surface. That makes one area cooler than the surface around it. The area gives off less light. From Earth, that area looks dark. We call it a ***sunspot.***

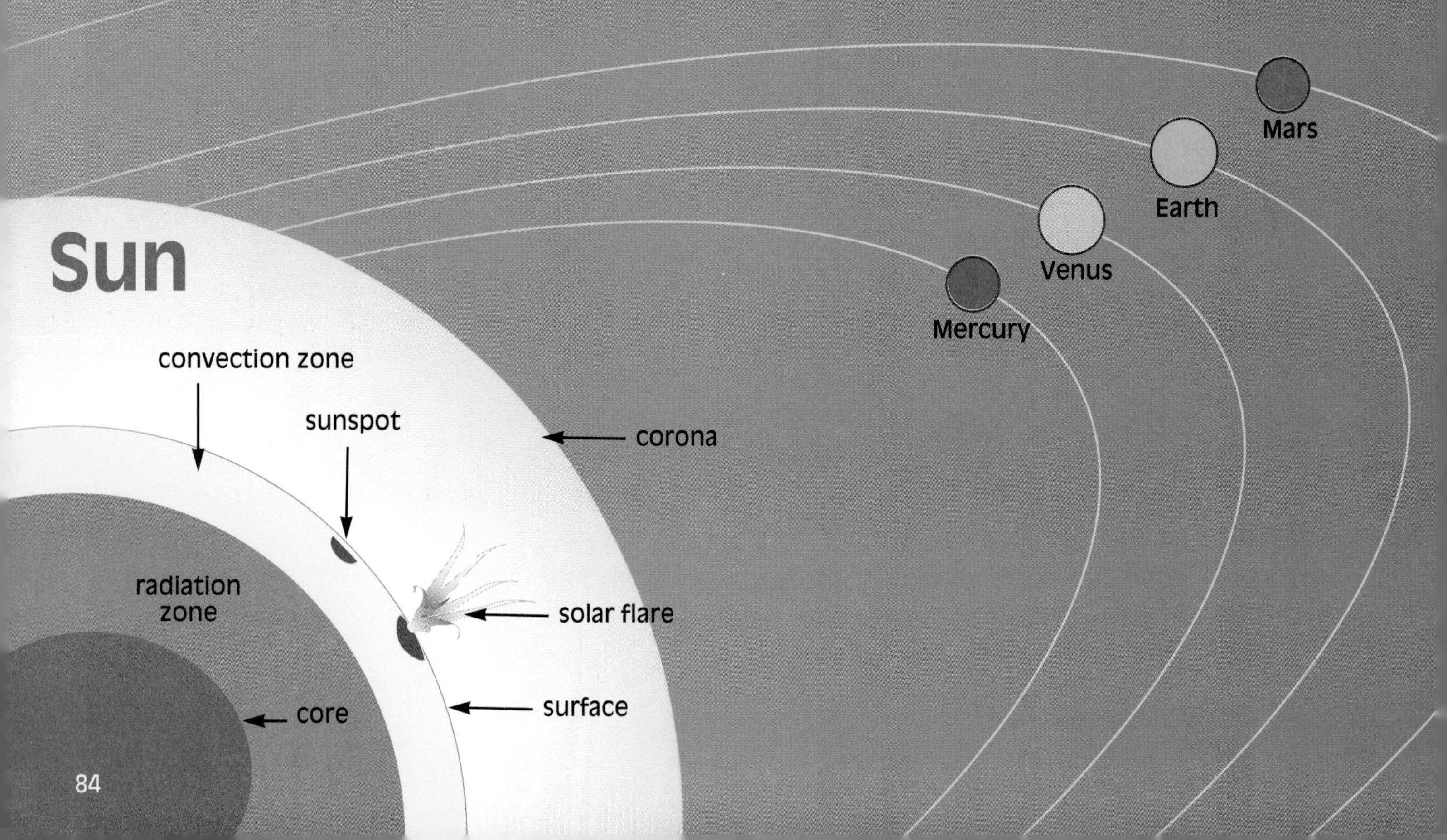

The Solar System

The sun and all the bodies that orbit around it make up the ***solar system.*** The solar system, about 4.65 trillion miles across, includes ***planets, moons, comets, asteroids,*** and other bits of rock and dust.

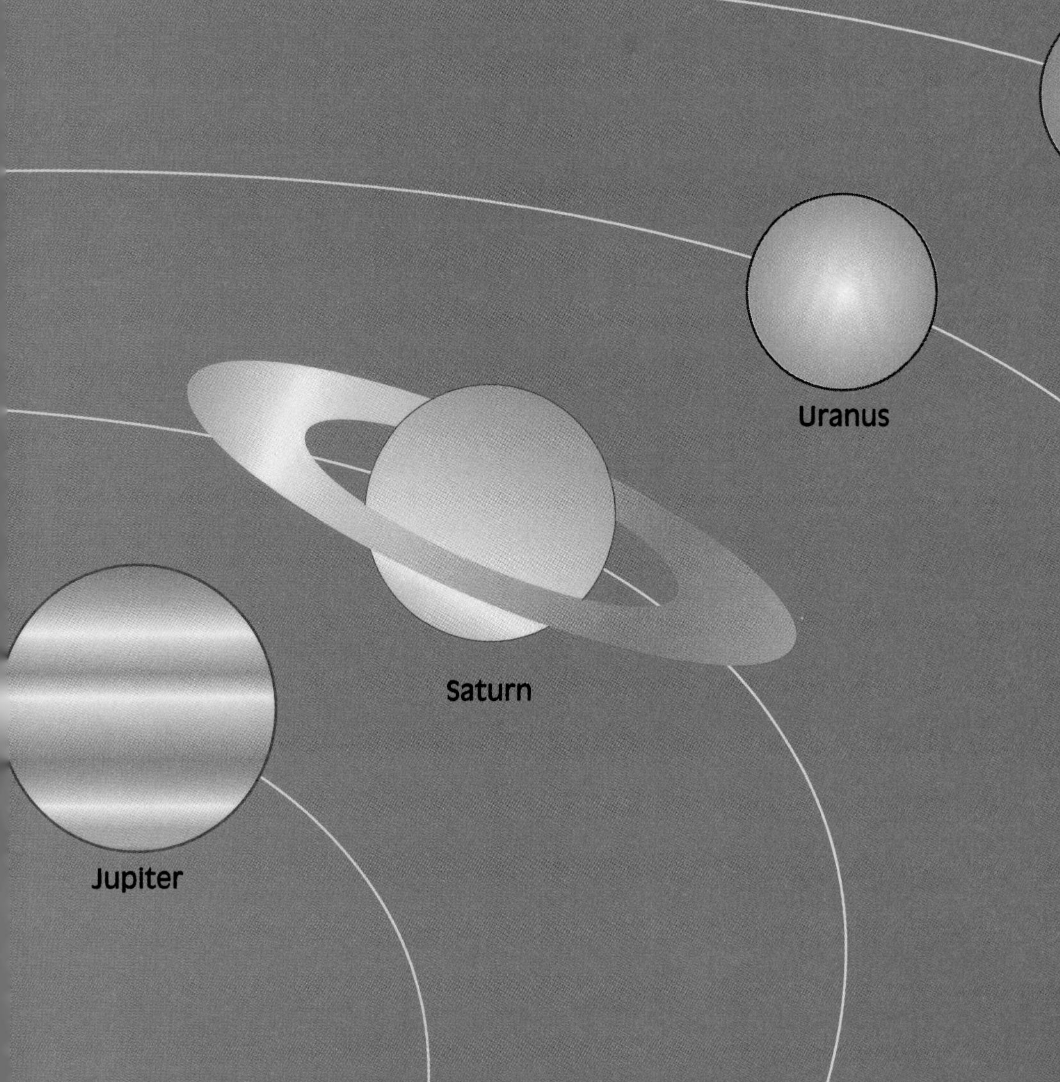

OUR SOLAR SYSTEM

Planet	Made of	Rings	Distance from sun (miles)	Moons	Diameter (miles)
Mercury	Rock	No	36 million	0	3,049
Venus	Rock	No	67 million	0	7,565
Earth	Rock	No	93 million	1	7,973
Mars	Rock	No	142 million	2	4,243
Jupiter	Liquid and gas	Yes	484 million	16	89,500
Saturn	Liquid and gas	Yes	887 million	18	75,000
Uranus	Rock	Yes	1.8 billion	15	32,125
Neptune	Frozen gas	Yes	2.8 billion	8	30,938
Pluto	Rock and ice	No	3.7 billion	1	1,875

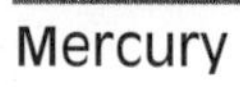
Mercury

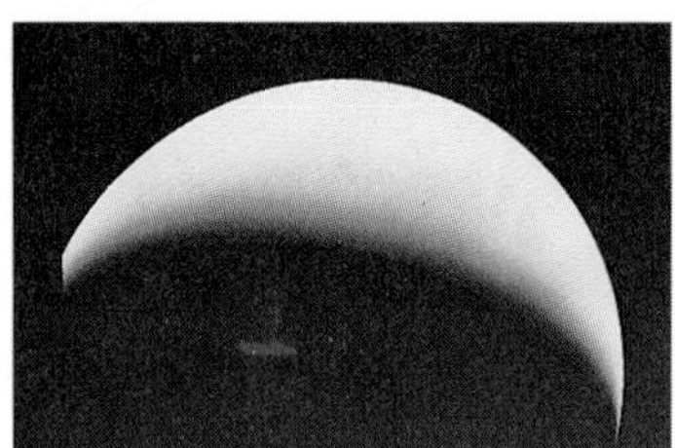
Venus

Earth

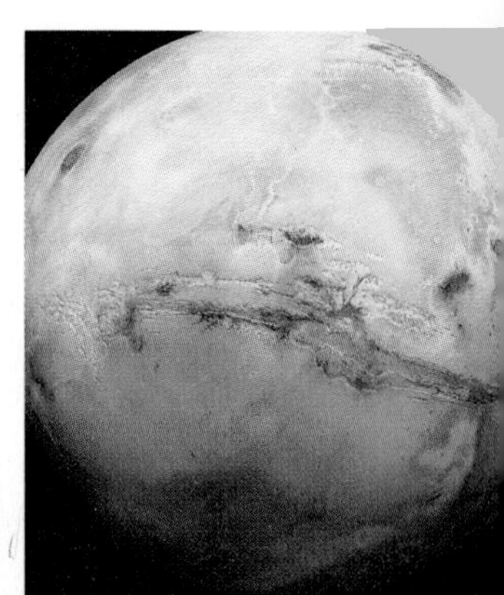
Mars

Length of year (in Earth time)	Length of day (in Earth time)	Gravity (compared to Earth)	Atmosphere	Temperature	Weight of object (100 pounds on Earth)
88 days	176 days	Less	Little, if any	425°C	37 pounds
225 days	116.7 days	Less	Carbon dioxide	450°C	88 pounds
365 days	24 hours	—	Nitrogen, oxygen, water	150°C	100 pounds
687 days	24.6 hours	Less	Carbon dioxide	120°C	37 pounds
11.9 years	9.9 hours	Greater	Hydrogen, methane	-1300°C	234 pounds
29.5 years	10.4 hours	Less	Hydrogen, methane	-1800°C	115 pounds
164.1 years	16 hours	Less	Hydrogen, methane	-2150°C	117 pounds
247 years	6.4 days	Greater	Hydrogen, methane	-2000°C	118 pounds
?	?	Less	Little, if any	-2500°C	?

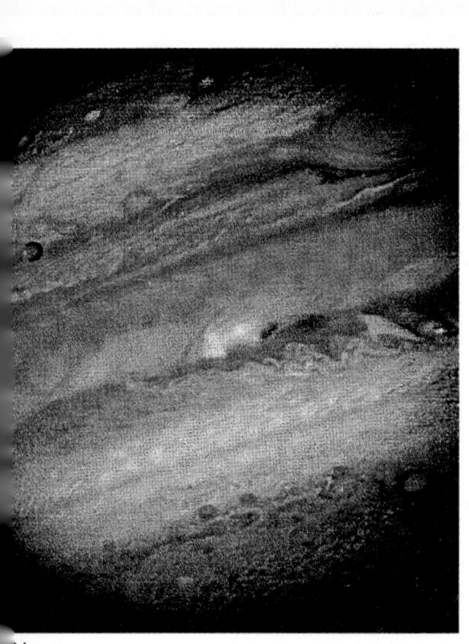

iter

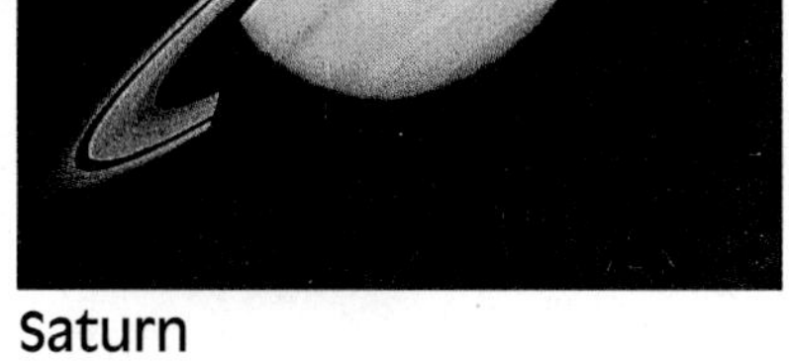

Saturn

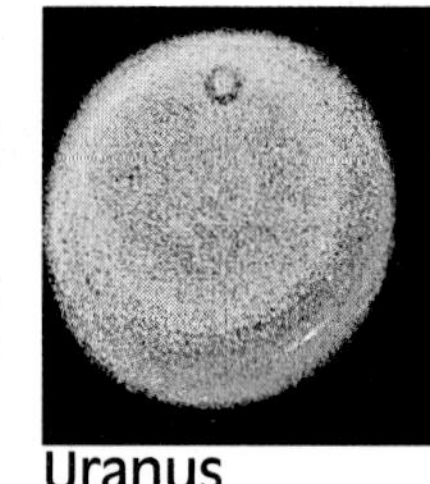

Uranus

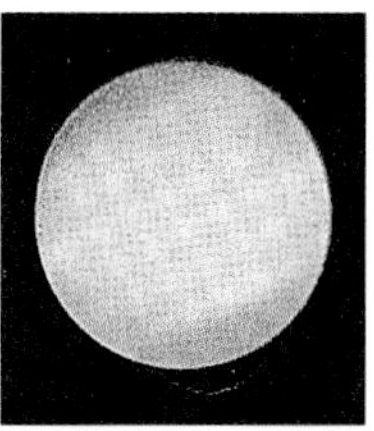

Neptune

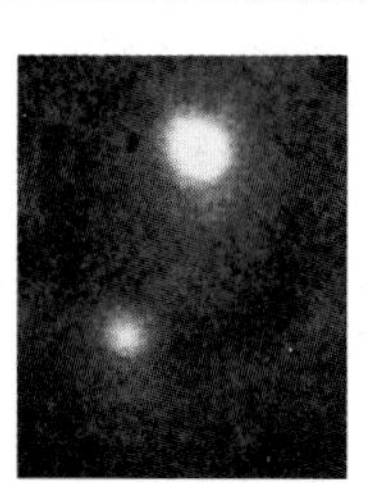

Pluto with moon

Old Theories about the Sun and Planets

The Sun at the Center

In 1543, Polish astronomer Nicolaus Copernicus published a theory that changed the way people thought about the world. Copernicus suggested that Earth revolves around the sun and that it does so in a circular orbit. Scholars of the day were convinced that the sun revolves around Earth. Unlike Copernicus, they had forgotten the ancient Greek astronomer Aristarchus who, around 300 B.C., argued that Earth revolves around the sun.

Elliptical Orbits

In 1609, German astronomer Johannes Kepler studied the observations of the sun's and the planets' positions made by Danish astronomer Tycho Brahe. Although Brahe's observations were made without a telescope (it had not been invented yet), Kepler used them to figure out that planets move in ***elliptical***, or oval-shaped, orbits.

Add Gravity

But why did the planets revolve around the sun? In 1665, English scientist Isaac Newton came up with the answer in his theory of universal gravitation. He said that all objects in the universe are attracted to each other by the force of gravity. The sun's gravity attracts the planets and other bodies in the solar system and prevents them from flying off into space.

Early astronomers thought that Saturn was the planet farthest from the sun. But in 1781, German-born musician and amateur astronomer William Herschel spied what seemed to be a comet (see p. 92) in his homemade telescope. When astronomers worked out the patterns of the object's orbit, they discovered it was a planet—Uranus. Neptune was discovered in 1846. Pluto remained unknown until 1930, when U.S. astronomer Clyde Tombaugh located the tiny planet among billions of stars in photographs.

The Earth and the Moon

Earth, the fifth-largest planet, orbits the sun once each year. As it revolves in its orbit, Earth also rotates on its ***axis,*** an imaginary straight line that runs through the north and south poles. The rotation causes the oppositie halves of Earth's surface to fall in and out of the sun's light in a regular pattern called ***days***. About 365 rotations, or days, are completed in the course of one revolution, or year. Earth's axis is not perpendicular to Earth's orbital plane, but tilts about twenty-three degrees. This means that at one side of its orbit, the northern half tilts toward the sun. This is spring and summer in the Northern Hemisphere and fall and winter in the Southern Hemisphere. When Earth moves so its northern half tilts away from the sun, we have fall and winter, and the Southern Hemisphere has spring and summer.

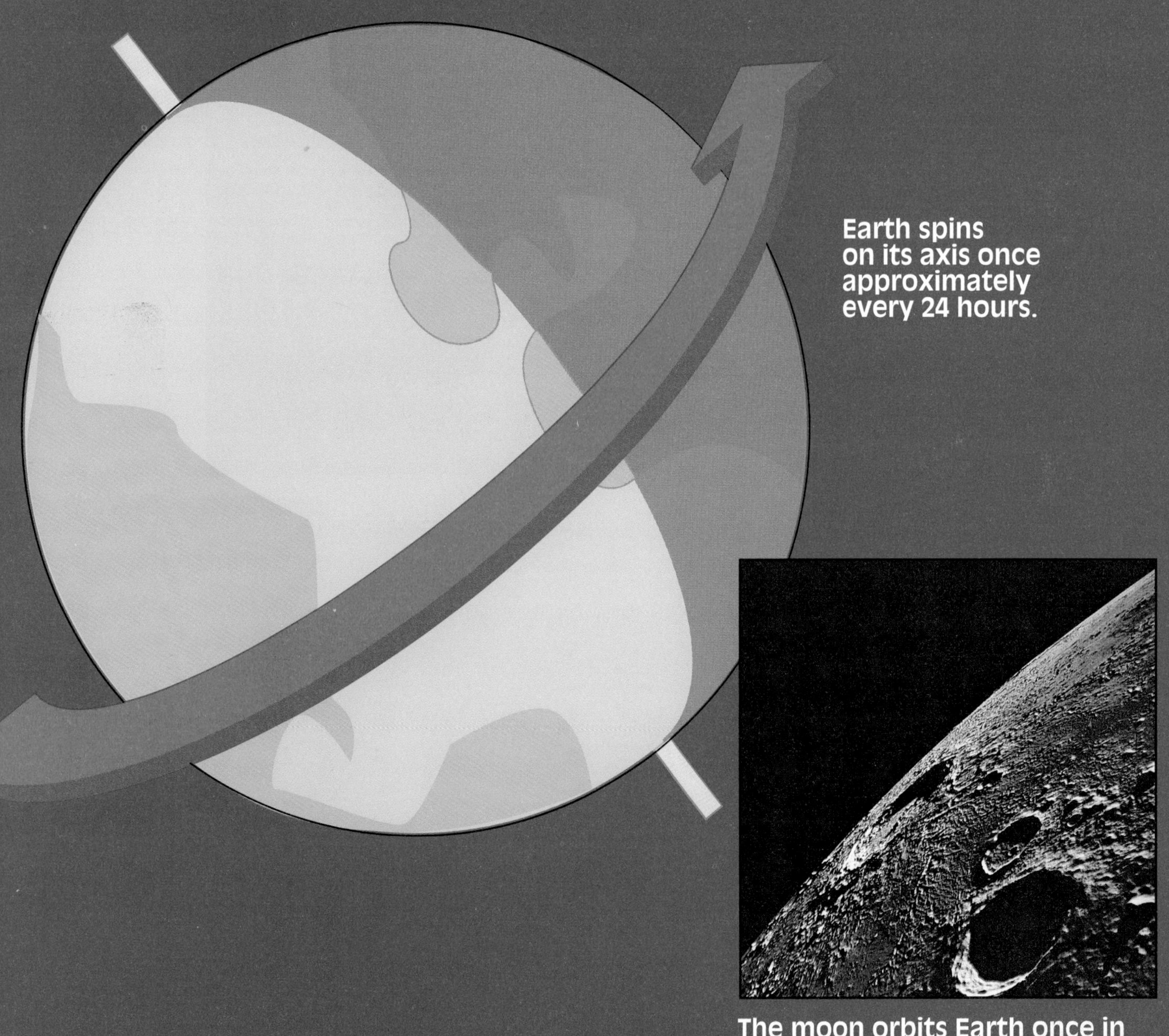

Earth spins on its axis once approximately every 24 hours.

The moon orbits Earth once in approximately 28 days.

Day of Darkness

Solar Eclipse

area of total eclipse
sun
Earth
moon
area of partial eclipse

solar eclipse

A satellite is an object that orbits another body in space. Earth is a satellite of the sun. The moon is a satellite of Earth. Many human-made satellites also orbit Earth. These satellites are used in telecommunications, security, weather science, and even in spying.

The moon is Earth's only natural ***satellite***. It is a giant rock that follows an elliptical (oval-shaped) orbit. The point of its orbit when the moon is closest to Earth is its ***perigee.*** The farthest point is its ***apogee***. The moon takes 27 1/3 days, about one month, to make one revolution around the earth. But only one face, or side, of the moon is ever visible from Earth. That's because Earth's gravity affects the moon's rotation in such a way that the moon completes precisely one rotation for every revolution around Earth. In fact, the pull of Earth's gravity is so strong that the moon's surface bulges toward our planet.

The moon is a satellite of Earth.

Phases of the Moon

The moon doesn't shine by its own light in the night sky. It reflects the sun's light toward Earth. Depending on where the moon is in its orbit, we see varying portions of the illuminated half of the moon. The different shapes of the moon are called ***lunar phases***. (The word ***lunar*** comes from ***lune***, the Latin word for ***moon***.)

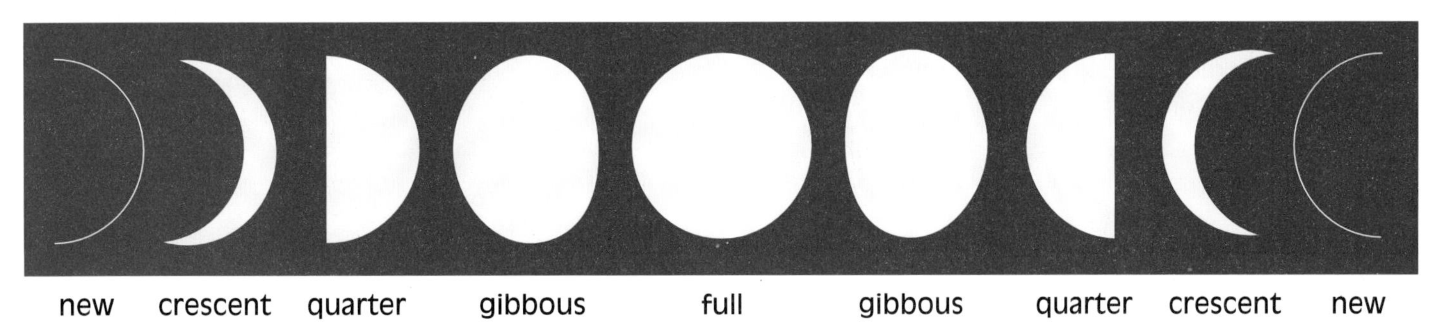

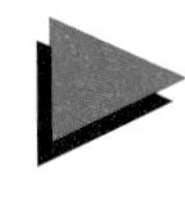

A **blue moon** *is a second full moon within one calendar month. A* **harvest moon** *is the full moon that occurs closest to the autumnal equinox. A* **hunter's moon** *is the next full moon after the harvest moon.*

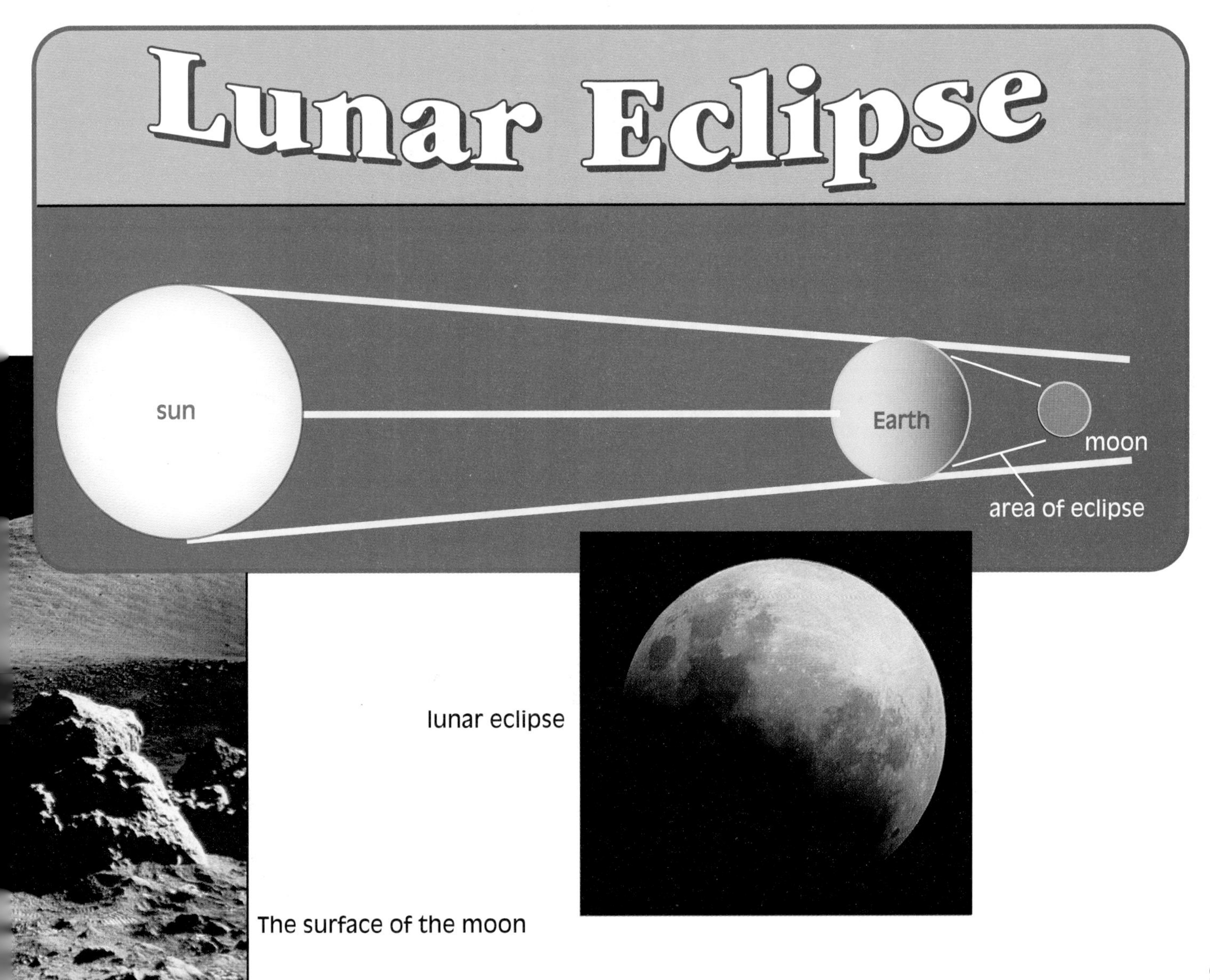

lunar eclipse

The surface of the moon

5 Asteroids, Meteors, and Comets

Asteroids

Asteroids are tiny planetlike bodies that revolve around the sun. Many asteroids are concentrated in the ***asteroid belt*** between Mars and Jupiter.

Meteors

Meteors are small rocklike objects that move from outer space into Earth's atmosphere. Friction with the atmosphere causes them to become white hot, and most meteors burn up before they reach Earth's surface. Meteors are also called ***shooting stars***.

Meteors that don't burn up completely can fall through the atmosphere and land on Earth. They are called ***meteorites***.

Comets

Comets are frozen masses of dust and ice that orbit the sun. When a comet passes near the sun, it turns into a bright ball with a blazing tail. When a comet approaches the sun, it appears brighter and brighter, and then fades as it passes by on its orbit.

Comets are rarely seen. Their orbits through space often take hundreds or thousands of years to complete.

comet with tail

Famous Comets

Halley's Comet	Donati's Comet
Encke's Comet	Schwassmann-Wachmann Comet
Great Comet of 1811	Arend-Roland Comet
Pons-Winnecke Comet	Ikeya-Seki Comet
Niela's Comet	Kouhoutek's Comet
Great Comet of 1843	Comet West

Space Exploration

Why Explore Space?

The current U.S. space shuttle program is designed to test everything from animal behavior to space diapers in a weightless environment. Meanwhile, satellites without passengers carry out many practical and exploratory tasks—photographing the earth for accurate mapping, observing weather patterns, and helping telecommunications. Communications satellites allow us to see television pictures from China or Australia while talking on the telephone to people in Argentina or Finland.

Chronology of Space Exploration

Year	Craft	Description
1957	Sputnik I	Soviets launch first unmanned artificial satellite.
1961	Vostok I	Soviet cosmonaut Yuri Gargarin travels in space for 1 hour, 48 minutes.
	Mercury 4	In the first U.S. manned space flight, Alan Shepard flies in a suborbital path for 15 minutes.
1962	Mercury 6	U.S. astronaut John Glenn circles earth 3 times in 4 hours, 55 minutes.
1963	Mercury 9	U.S. astronaut Gordon Cooper orbits earth 22 times in 1 1/2 days.
1965	Gemini 3-6	U.S. astronauts maneuver spacecraft to change flight paths (*Gemini 3*), meet up in space (*Gemini 6* and *7*), and set new record for time spent in space — 13 days, 18 hours.
1968	Apollo 8	U.S. spacecraft orbits moon 10 times.
1969	Apollo 11	U.S. astronauts Neil Armstrong, "Buzz" Aldrin, and Michael Collins land on the moon.
1971	Apollo 17	Last U.S. mission to the moon.
1973	Skylab I	First *Skylab* crew is sent into space for 28 days.
1975	Apollo Soyuz	First U.S.- Soviet collaborative flight.
1981	Space Shuttle	The shuttle *Columbia* successfully orbits and returns to Earth to be refueled and flown again.
1983	Space Shuttle	Sally Ride aboard *Challenger* becomes the first U.S. woman astronaut.
1986	Space Shuttle	*Challenger* explodes after takeoff, killing all its crew, including schoolteacher Christa McAuliffe.
1989	Space Shuttle	*Atlantis* crew releases unmanned *Magellan* probe to explore Venus.

THE PHYSICAL WORLD

1 Matter

Matter is anything that has ***volume*** and ***weight.***

Volume

States of Matter

The amount of space something takes up is its ***volume***. Solid things—tables, chairs, desks, windows, walls, and floors—take up space and so have volume. Liquids, such as water in an aquarium or juice in a carton, also have volume. What about the air around you? You can't see it, but you know it's there. Air is made up of gases, which also have volume.

Matter can be ***solid, liquid***, or ***gas***. These three forms are called the ***states of matter***.

Temperature and Volume

The volume of matter changes with temperature. If a solid, liquid, or gas is heated, its volume will usually ***increase***. If a solid, liquid, or gas is cooled, its volume will usually ***decrease***.

Gases change volume more easily than solids. They fill whatever container they are in. Like liquids and solids, gases usually expand when heated under constant pressure. The cooler the gas, usually the smaller its volume.

Mass

Think about a small block of wood and a block the same size made of iron. If you kick the block of wood, it will go flying. But kick the block of iron and you'll probably stub your toe. That's because the iron has more ***mass***, or amount of matter, in it.

Set both blocks on a table. Neither will move without a push or pull from you. The force you have to overcome to get the blocks moving is called ***inertia***. You will need to put more effort into moving the iron block than the wood block because of its greater mass.

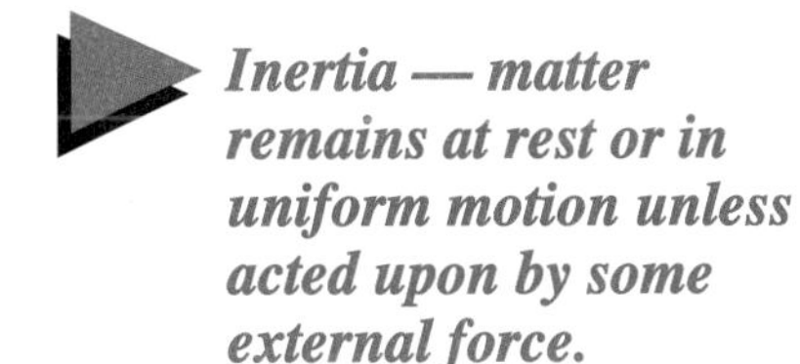

Inertia — matter remains at rest or in uniform motion unless acted upon by some external force.

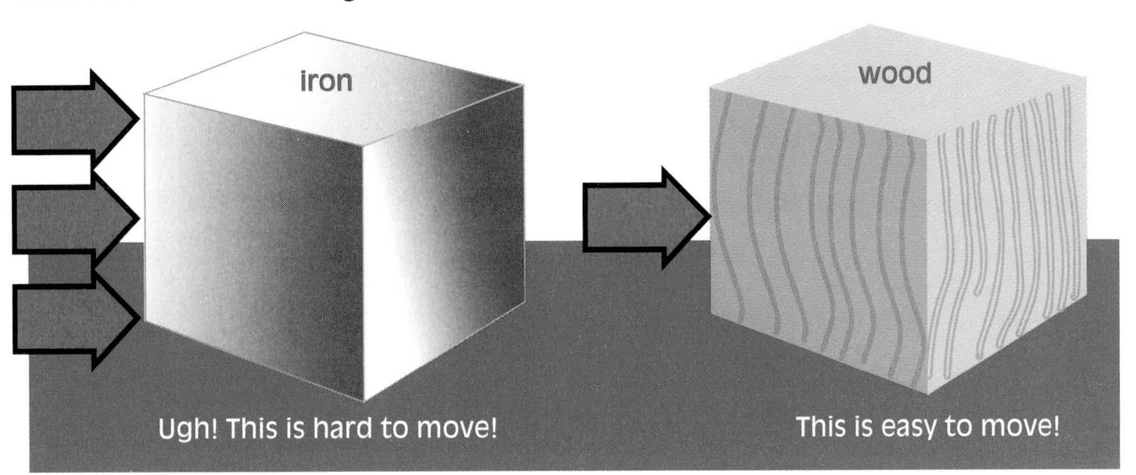

Weight

Weight shows gravity's pull on an object. If you put the iron and the wood blocks on a scale, the iron will weigh more than the wood. That's because Earth's gravity is pulling on more "stuff" in the iron block than in the wood block. On the moon, both objects will weigh less than they do on Earth, because the moon's gravitational force, or pull, is less than that of Earth. If you kicked the iron block on the moon, it, too, would go flying, much like the wood block on Earth.

Weight can change from place to place, but the mass, or amount of "stuff" in an object, stays the same everywhere.

What's in it?

Substance

A simple, pure thing with a single name, such as glucose. (A substance can be a compound.)

Mixture

Two or more things that maintain their separate identities when mixed together, such as salt and pepper.

Solution

A liquid mixture that may look like one thing but that remains separate, such as salt water.

Element

A substance, such as copper, that cannot be broken down into a simpler substance by ordinary chemical means.

Compound

A pure substance formed by two or more elements reacting to form another new substance that is chemically different from any of the single elements. Table salt is a compound called ***sodium chloride*** (see also Elements, p. 98).

Density

Density describes the mass of an object divided by its volume. A container of feathers is lighter than the same container filled with bricks.
(See also Changing Matter, p. 100.)

feathers have less density than bricks

gases are less dense than solids or liquids

Why do objects sink or float? Objects that are denser than their surroundings sink. Objects that are less dense than their surroundings float.

Conservation of Matter and Energy

Matter cannot be created or destroyed in ordinary chemical and physical processes—although it can change from one form to another. Scientists call this the ***law of conservation of mass***. Energy can also change forms, but it cannot be created or destroyed. This is the ***law of conservation of energy.***

In 1905, German-born scientist Albert Einstein showed that matter can be ***changed*** into energy and that energy can be changed into matter. These changes take place only in extreme conditions, like nuclear reactions. But, ***the total of matter and energy will always be the same because matter and energy cannot be created or destroyed***. Einstein's theory is called the ***law of conservation of mass-energy.***

Atoms and Molecules

Atoms

Atoms are tiny building blocks of matter. They are too small to be seen even with a microscope. Even the point of a pin is enormous compared with an atom. But scientists have methods of studying atoms. They know how to weigh them, measure them, and tell them apart.

Parts of an Atom

As tiny as they are, atoms are made up of even tinier parts. At the center of an atom is a core called a ***nucleus***. The nucleus is made up of particles called ***protons*** and ***neutrons.*** Protons have a positive electrical charge. Neutrons have no charge. Scientists in the early twentieth century thought that electrons orbit around the nucleus like the planets of our solar system orbit around the sun. Now they know that electrons do not move in regular "orbits." They move around the nucleus in ***electron shells,*** or energy levels, each of which can hold a certain number of electrons. Electrons have a negative charge. They are attracted to the positively charged protons in the nucleus. This attraction keeps the electrons in orbit around the nucleus.

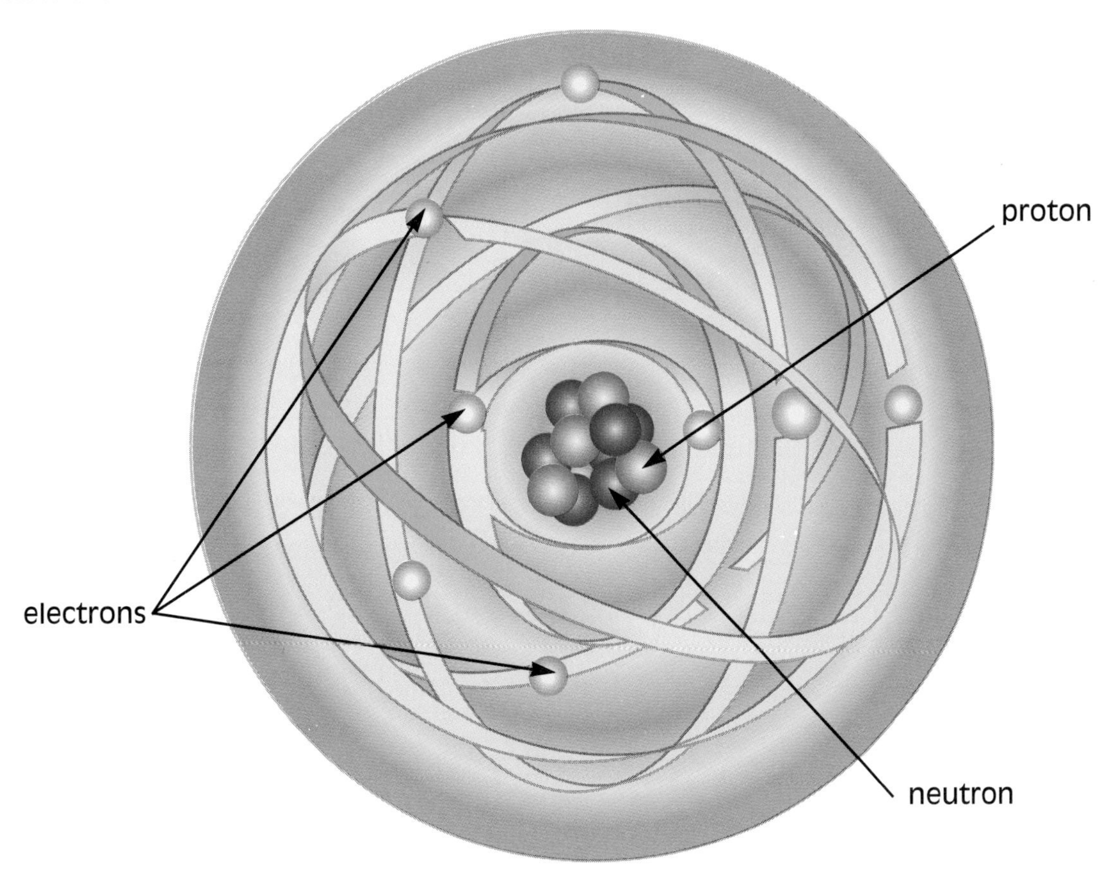

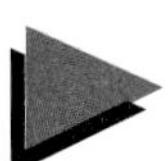

More than five million atoms can fit on the head of a pin.

Elements

All atoms are made up of protons, neutrons, and electrons, except for one type of hydrogen atom, which has no neutron. But not all atoms have the same numbers of these particles. So far, scientists have discovered 109 atoms with different combinations of protons, neutrons, and electrons and different properties. These 109 substances are called ***elements.***

A Table of Elements

In 1869, Russian chemist Dmitri Mendeleev created the periodic table of elements. He organized the known elements by their atomic weights and chemical properties. These properties depend on the number and arrangement of electrons in atoms of the elements. Mendeleev's work predicted the existence and weights of several elements that were not yet known. A few elements' weight order differs from their atomic number order, and by 1913, English scientist Henry G. J. Moseley had rearranged the elements according to their atomic numbers. (Atomic numbers are determined by the number of protons in the nucleus of an atom of the element, which is also the number of electrons.)

1 H						
3 Li	4 Be					
11 Na	12 Mg					
19 K	20 Ca	21 Sc	22 Ti	23 V	24 Cr	25 M
37 Rb	38 Sr	39 Y	40 Zr	41 Nb	42 Mo	43 T
55 Cs	56 Ba	57 to 71	72 Hf	73 Ta	74 W	75 R
87 Fr	88 Ra	89 to 103	104 Unq	105 Unp	106 Unh	107 Un

57 La	58 Ce	59 Pr	60 Nd	61 Pr
89 Ac	90 Th	91 Pa	92 U	93 N

The science of matter is called **chemistry.** *Scientists who study chemistry are called* **chemists.** *The science of how things work and move is* **physics.** *Scientists who study physics are called* **physicists.**

Key

22 ← Atomic number

Ti ← Symbol of element

										2 He
					5 B	6 C	7 N	8 O	9 F	10 Ne
					13 Al	14 Si	15 P	16 S	17 Cl	18 Ar
6 Fe	27 Co	28 Ni	29 Cu	30 Zn	31 Ga	32 Ge	33 As	34 Se	35 Br	36 Kr
4 Ru	45 Rh	46 Pd	47 Ag	48 Cd	49 In	50 Sn	51 Sb	52 Te	53 I	54 Xe
6 Os	77 Ir	78 Pt	79 Au	80 Hg	81 Tl	82 Pb	83 Bi	84 Po	85 At	86 Rn
08 Jno	109 Une									
2 Sm	63 Eu	64 Gd	65 Tb	66 Dy	67 Ho	68 Er	69 Tm	70 Yb	71 Lu	
4 Pu	95 Am	96 Cm	97 Bk	98 Cf	99 Es	100 Fm	101 Md	102 No	103 Lw	

Subatomic Particles

Protons, neutrons, and electrons are ***subatomic particles***. They are smaller than atoms. For years, scientists thought that these three particles were the building blocks of atoms. They now believe that two kinds of subatomic particles, ***quarks*** and ***leptons***, make up atoms—and all matter.

Quarks are charged particles that make up neutrons and protons in the nucleus of an atom. There are six kinds, or flavors, of quarks: up, down, strange, charmed, truth, and beauty. "Truth" and "beauty" are also called "top" and "bottom" by some scientists.

There are six lepton particles and they are usually grouped in pairs.

Molecules

Atoms often group together with other atoms into ***molecules.*** Some molecules are made up of the same element. Others are made up of combinations of elements. Molecules can contain anywhere from two to thousands of atoms. Any substance made up of two or more different elements is called a ***compound***. A molecule that includes carbon atoms is called an ***organic compound***. That means all living things (and fossils of once living things)—***organisms***—contain carbon.

Changing Matter

Matter can be changed in two ways: physically and chemically.

Physical changes do not involve one substance changing into another. Water, for example, can change from gas (water vapor), to liquid water, to solid (ice), but the water molecules don't change.

Chemical changes occur when a substance is changed into something else. When paper is burned, for example, the molecules are altered, transforming the paper into gases and ash. The burning paper also gives off heat and light energy (see p. 101).

Most chemical changes involve the outer electron layers of atoms. But ***nuclear*** changes affect the nucleus. Sometimes a nucleus breaks apart. This change is called ***nuclear fission***. A nucleus can also change by joining together with another nucleus. This change is called ***nuclear fusion.***

Acids are substances that have a sour taste. Weak acids, such as citric acid in lemons and limes, or acetic acid in vinegars, are found naturally in the foods we eat. Even cola drinks are acids.

Hydrochloric acid is in our bodies to help start such natural processes as digestion and to fight germs and bacteria. Other acids are made by humans for industrial use. Very strong acids, such as sulphuric and nitric acids, ***corrode***, or eat away, tough materials, including wood, rubber, and metal.

Scientists measure the strength of acids on a ***pH scale***. Pure water — which has no acid content at all — has a pH of *7*.

The pH scale is also used to measure ***bases***. Bases are also called ***alkalis***. These substances feel soapy. Like acids, they can corrode other materials and they are measured on a pH scale.

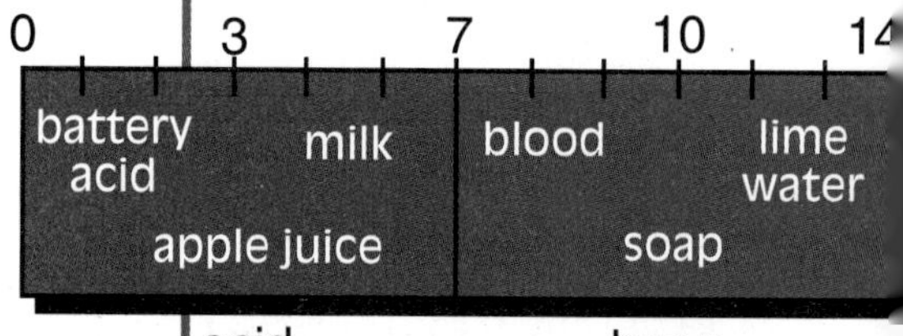

Litmus paper is a special paper used to measure the pH of substances. If litmus paper turns ***red***, the substance is an acid. If the paper turns ***blue***, the substance is a base. The intensity of the color also shows if the acid or base is strong or weak.

Energy, Motion, and Work

Energy is the ability to do work. ***Work*** means moving something against resistance. If you want to lift a heavy box, you use energy to move the box against the force of gravity (resistance) from the floor to your desktop.

Energy

Energy takes many forms—radiant (visible and infrared light), sound, and electric, to name a few. Different forms of energy can be changed from one kind to another. But energy itself can't be created or destroyed in ordinary chemical or physical processes.

All energy is either ***potential*** or ***kinetic***. When an object has potential energy, it means that if the object were set into motion, it would acquire kinetic energy. Kinetic energy is "active" energy. Objects with kinetic energy are moving. Potential energy can always be changed into kinetic energy and back again into potential energy.

Energy is locked into the nucleus of every atom. This energy is called nuclear energy. Nuclear energy is released by splitting atoms in a process called nuclear fission. In a power plant, the nuclei of uranium atoms are split to create electricity. Uranium is used because it is unstable and the potential energy of uranium nuclei is easily changed to kinetic energy.

Measuring Energy

All forms of energy can be changed into ***heat*** (see page 107-113).

All forms of energy, including heat, are measured in ***joules.*** A joule (J) is the kinetic energy possessed by a two-kilogram mass moving at a velocity of one meter per second.

Heat is sometimes measured by its ability to raise the temperature of water. Both ***calories*** and ***British thermal units*** (BTUs) show this measurement (see Measuring Heat, p. 110).

Let the Force Be With You

A ***force*** is a push or a pull on an object. A push or a pull can set a st object into motion. A push or pull against a moving object can stop it Force can also make a moving obj travel faster or in a different directi

Force is measured in ***newtons*** (n

Motion

Motion is the act or process of moving from one place to another. Motion is explained by three basic laws.

The First Law of Motion

An object at rest tends to remain at rest and an object in motion tends to continue moving in a straight line at constant speed. The tendency of objects to remain at rest or in motion is called ***inertia***. A ***force*** is needed to overcome inertia. Force—a push or pull—will start a ball at rest rolling. Force, such as friction, can also stop a rolling ball.

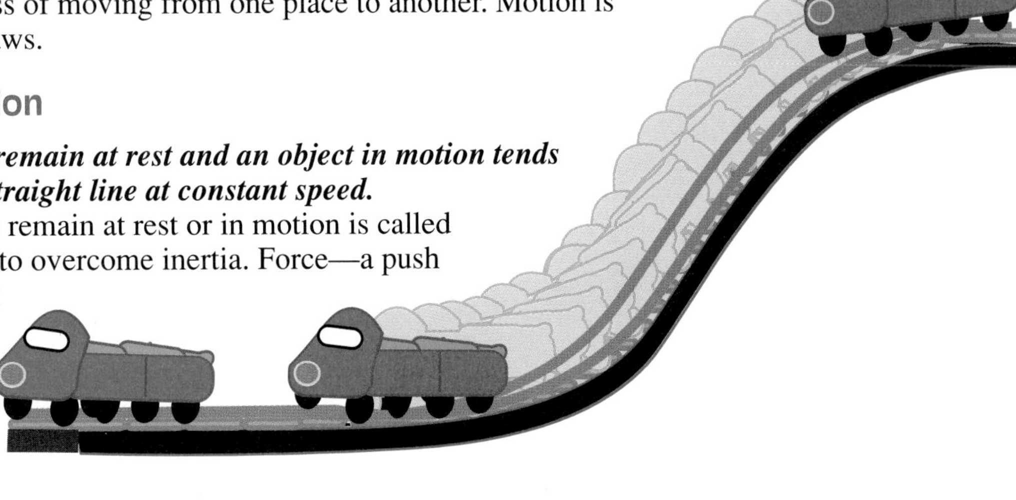

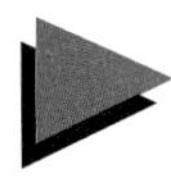

The laws of motion, known as Newton's laws, are named for Sir Isaac Newton, the English scientist who stated them in 1687.

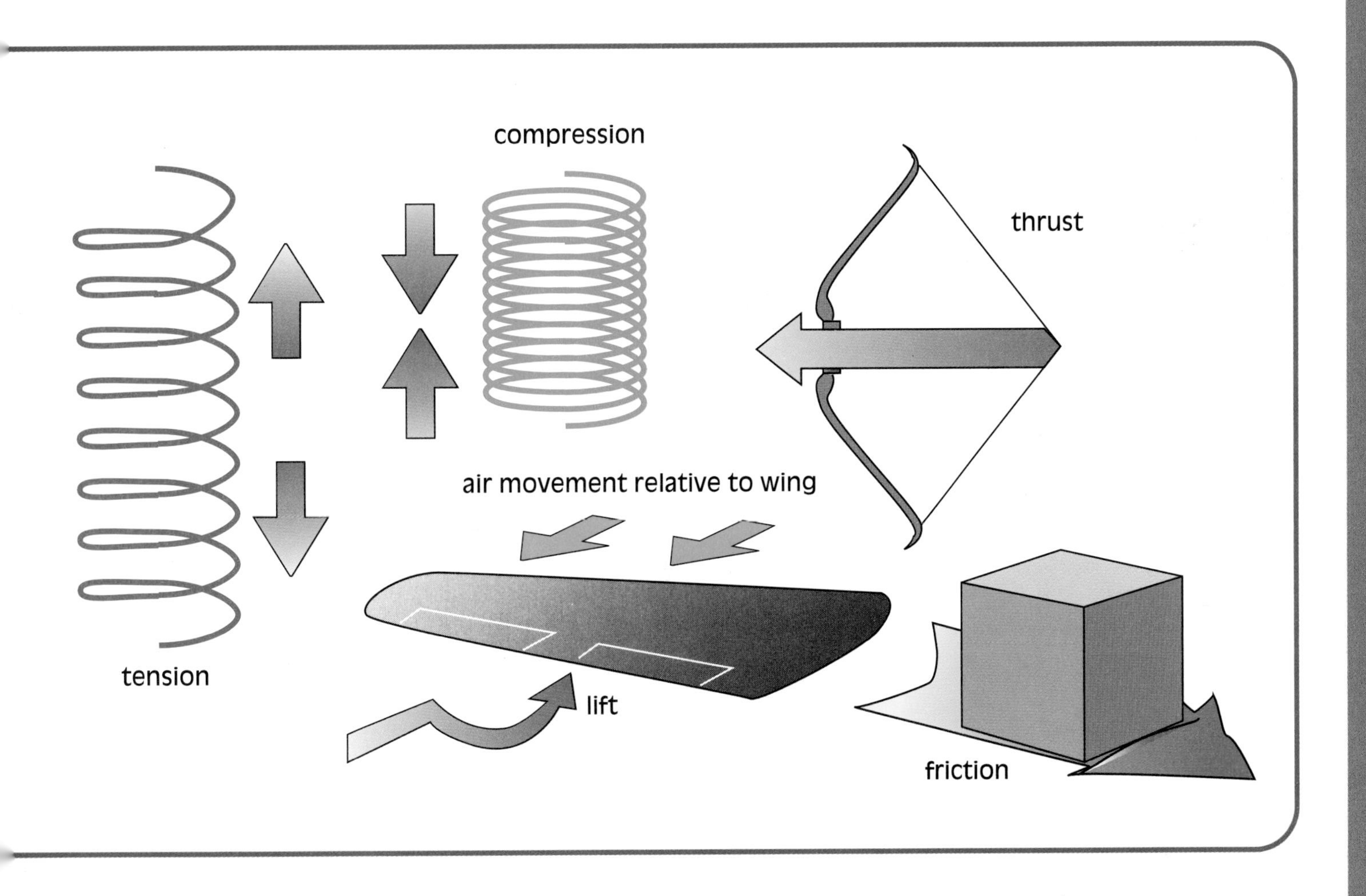

The Second Law of Motion

Force equals mass times acceleration.
This law says that if you apply force to an object, you will either change its speed, change its direction, or both. The harder a pitcher throws, for example, the faster the ball will travel. The law also says that you will have to use more force to move or stop a massive object, like a refrigerator, than to move or stop a picnic cooler.

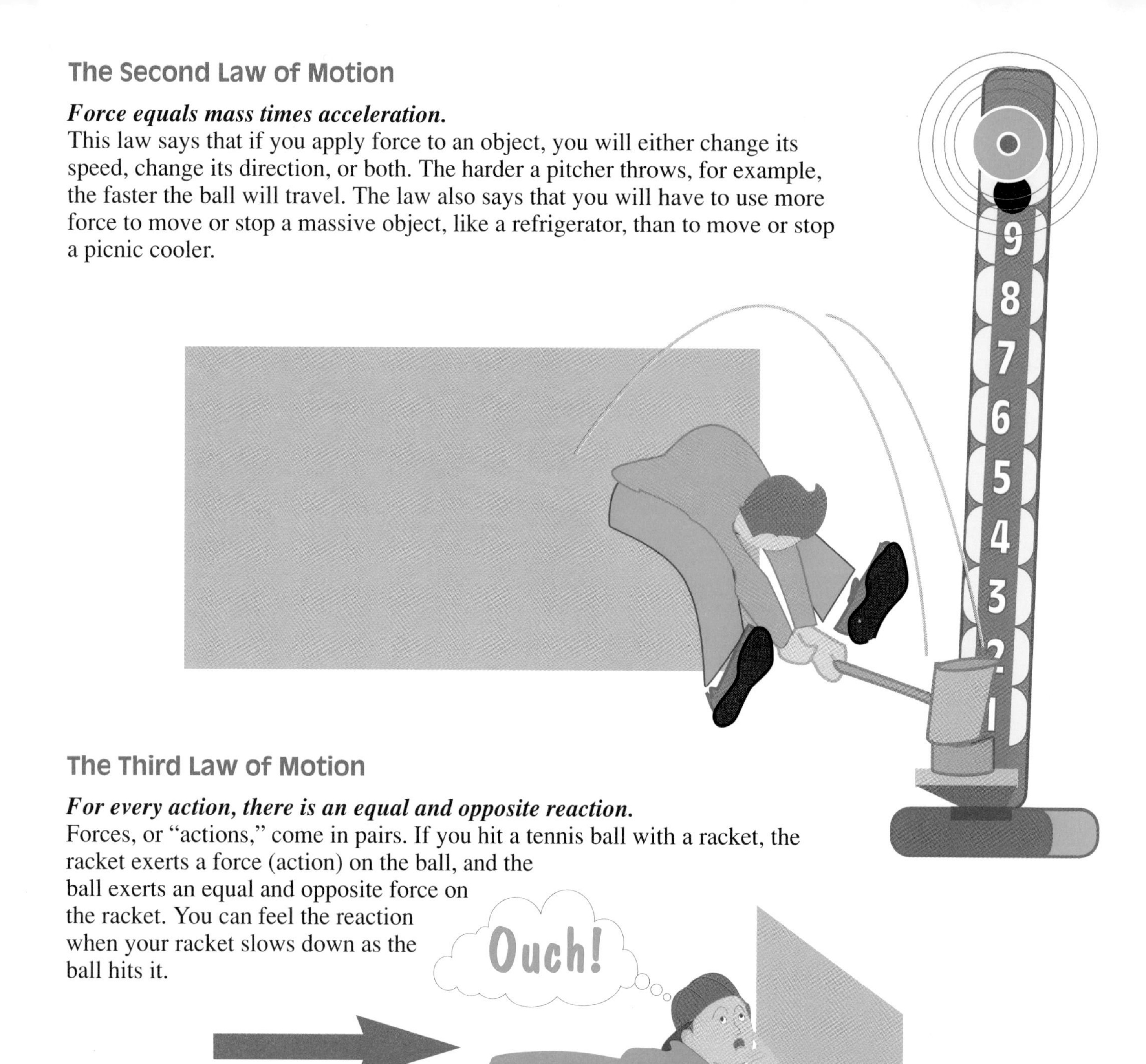

The Third Law of Motion

For every action, there is an equal and opposite reaction.
Forces, or "actions," come in pairs. If you hit a tennis ball with a racket, the racket exerts a force (action) on the ball, and the ball exerts an equal and opposite force on the racket. You can feel the reaction when your racket slows down as the ball hits it.

Work

When a force causes an object to move, ***work*** is done. The ***amount*** of work is measured by multiplying the force times the distance an object moves.

work = force x distance an object moves in the direction of the force

The work needed to lift a twelve-pound sack of pennies two feet is calculated:

work = 12 pounds x 2 feet
work = 24 foot-pounds

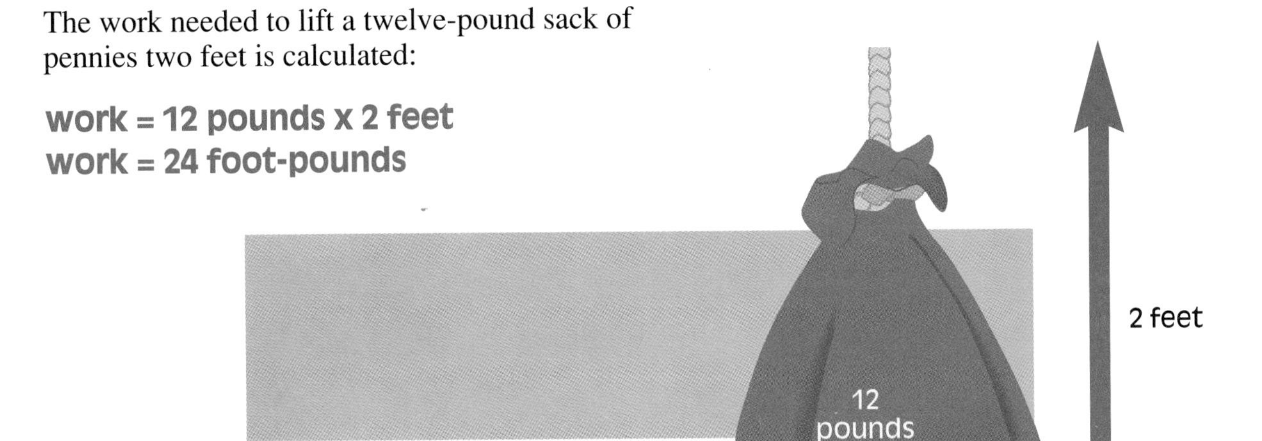

It takes a force of 0.1 pounds to make a book slide on a surface. How much work is done when you slide a book four feet?

work = 0.1 pounds x 4 feet
work = .4 foot-pounds

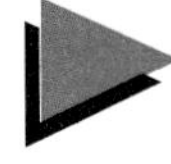

Since distance is commonly measured in feet and weight (resistance) *in pounds,* work *is measured in a combination of feet and pounds called* foot-pounds. *Scientists measure work in joules (see Measuring Energy, p. 102).*

Simple Machines

Imagine riding your bike straight up a steep hill. Now imagine climbing the same height along a gently rising path that zigzags across the hillside.

The zigzag path would certainly make an easier climb, but since the same mass is moved the same net distance against gravity, the work is the same. Still, the gentle route requires less force, so it offers a ***mechanical advantage*** over the straight climb.

The advantages of reducing the force necessary to do work are well known. The tools of early humans made use of the simple machines we still use today to make work easier. These machines all offer a mechanical advantage.

A rotary force on the screw will move the load up. If the force moves the screw in the other direction, the load will move down.

force

screw

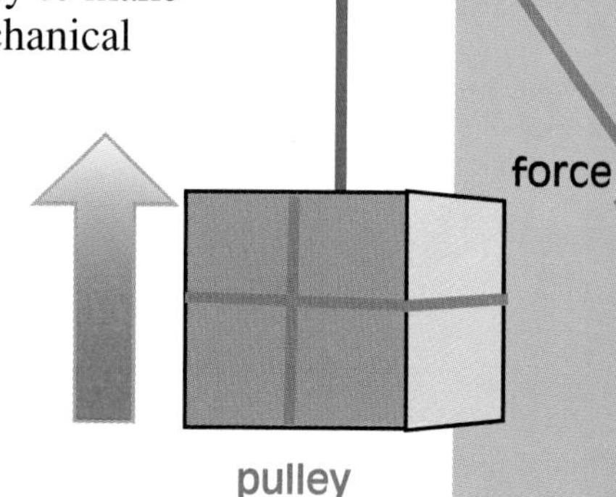

A downward force on the pulley cable will move the load up.

pulley

wheel and axel

force

A downward force on one end of the lever will move the other end up.

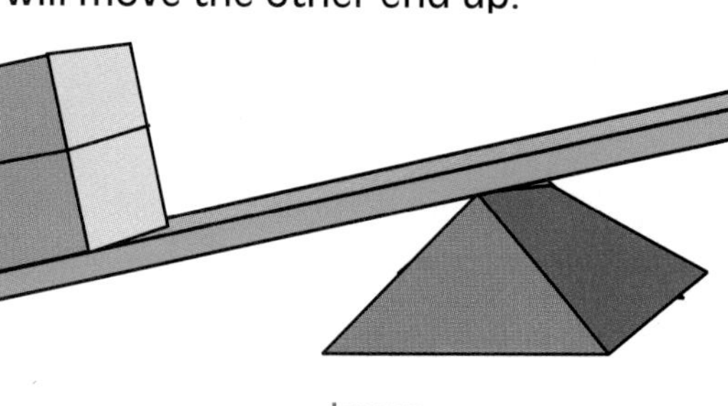

lever

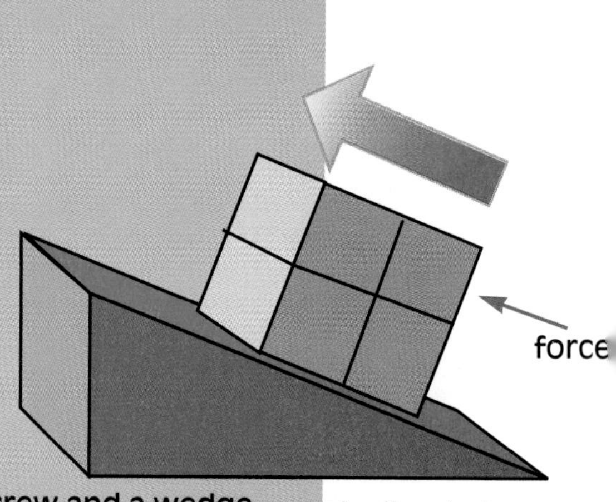

inclined plane

A screw and a wedge are forms of inclined planes. Force is used to move loads up gradually rather than straight up, making use of mechanical advantage.

POWER to the PEOPLE

Machines help people do work faster. The speed of work is called ***power.*** Power is equal to the work done divided by the time it takes to do it.

$$\text{power} = \frac{\text{work}}{\text{time}}$$

Heat and Temperature

Heat

We use heat every day to warm our homes and cook our food.
Heat is a form of kinetic energy (see p. 101). Heat is created when atoms and molecules in a substance vibrate, or move backward and forward very rapidly. As a substance gets hotter, its molecules move faster and faster.

Temperature

Temperature tells how hot or cold something is. Temperature also tells you whether heat is flowing into a substance or flowing out of it.

Measuring Temperature

Temperature is measured with a thermometer, an instrument containing a substance that indicates the temperature of its surroundings. It does so by reaching the same temperature as its surroundings.

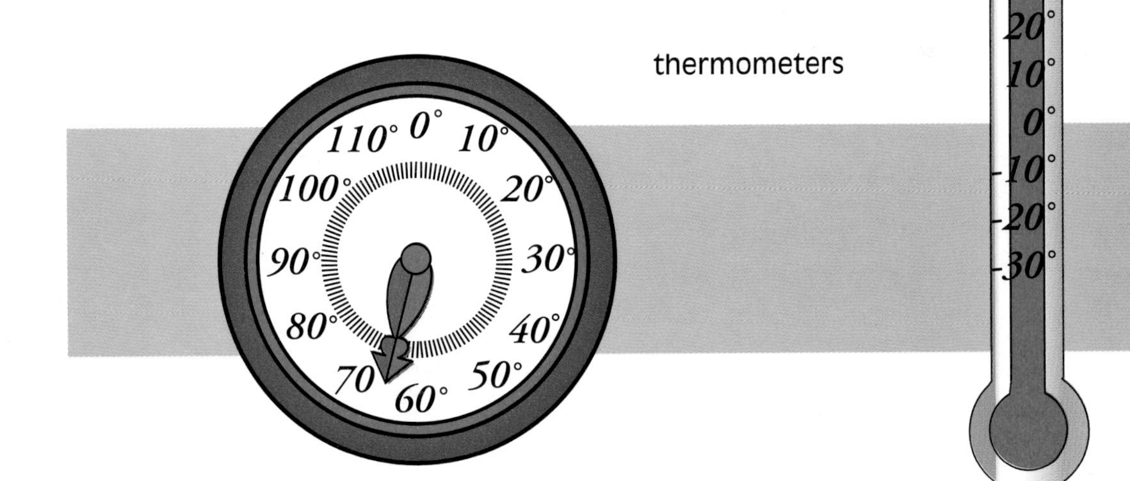

thermometers

A LUKEWARM EXPERIMENT

More than 300 years ago, the English thinker John Locke was convinced that the sense of touch could not be used to measure temperature. He devised an experiment to prove his point. You can try it yourself. Was Locke right?

You'll need:
three bowls
ice cubes
tap water

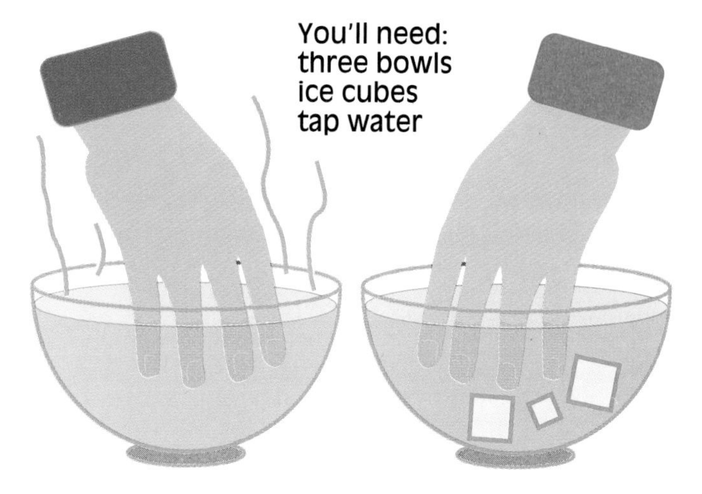

Fill one bowl with hot water. Fill another with cold water, and add ice to it. Fill the third bowl with lukewarm water.

Put one hand in the bowl of hot water while you put the other hand in the bowl of cold water. How do your hands feel?

Next, put both hands in the bowl of lukewarm water. Now how do your hands feel?

To the hand that has been in hot water, the lukewarm water feels cold. To the hand that has been in ice water, the lukewarm water feels warm.

Galileo's Thermometer

Italian scientist Galileo Galilei built one of the first thermometers. He used a bulb-shaped glass with a long, thin tube. The tube was open on one end. Then Galileo slightly heated the air inside the bulb and turned it upside down in a pan of water. When the air cooled, water rose part way up the narrow part of the tube. Then, when the surrounding temperature became hotter, the air in the bulb expanded, driving the liquid down in the tube. When the surrounding temperature became cooler, the water rose in the tube. The height of water in the tube would roughly indicate the temperature of the surroundings.

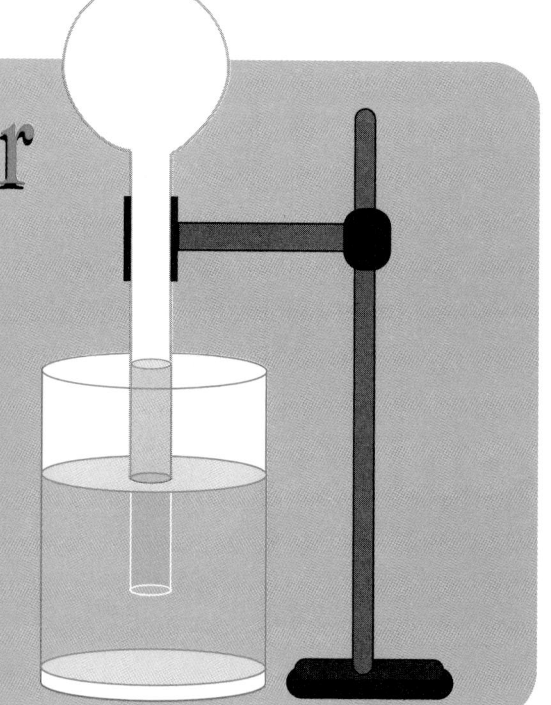

Fahrenheit and Celsius

The first accurate thermometer was invented by Gabriel Daniel Fahrenheit, a German physicist, in the early 1700s. Fahrenheit created a scale for his thermometer and named each unit of change a ***degree***. The temperature measured on the Fahrenheit scale is written as °F, which means "degrees Fahrenheit." On this scale, at standard sea level atmospheric pressure, water freezes at 32°F and boils at 212°F.

In 1742, Anders Celsius, a Swedish astronomer, created a new scale for measuring temperature. The Celsius scale put the freezing point of water at 0°C (zero degrees Celsius) and the boiling point at 100°C. The Celsius scale—also known as the centigrade scale—is the world standard for measuring temperature.

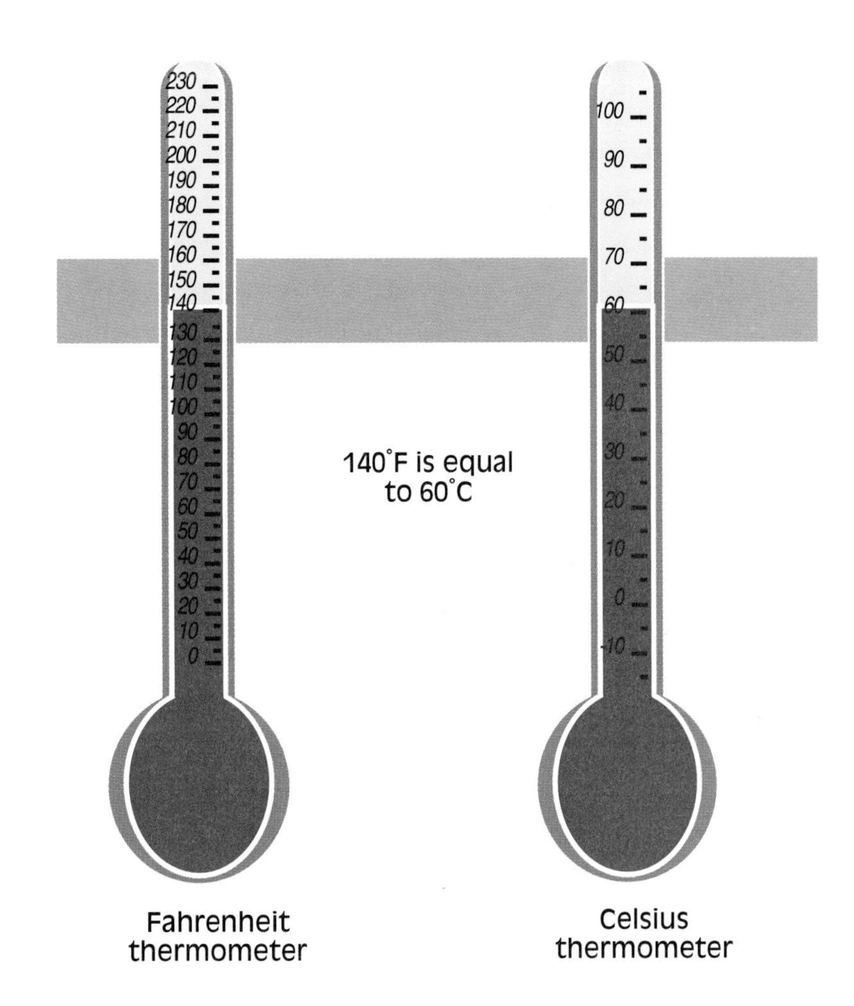

Fahrenheit to Celsius

To compare temperatures measured in Fahrenheit and Celsius, you need to convert the temperatures to one scale or the other.

To convert Fahrenheit to Celsius, subtract 32 from the Fahrenheit temperature. Then multiply the result by 5/9.

To convert Celsius to Fahrenheit, multiply the Celsius temperature by 9/5. Then add 32 to the result.

Fahrenheit/Centigrade Equivalents

°F	°C
0	-17.8
10	-12.2
20	-6.67
32	0
40	4.44
50	10
100	37.78
212	100

°C	°F
-50	-58
0	32
10	50
20	68
30	86
40	104
50	122
100	212

Kelvin

Scientists use another temperature scale called the Kelvin scale. It was devised around 1850 by the British scientist William T. Kelvin. On the Kelvin scale, one degree of change is the same as one degree of change on the Celsius scale. But the freezing point of water is measured at 273°K (degrees Kelvin) and the boiling point at 373°K. Zero degrees on the Kelvin scale means ***absolute zero***, or the point at which an object has no heat.

Three Laws of Thermodynamics

Since Kelvin's time, scientists have proved that absolute zero is impossible to reach. Nothing can ever be so cold that it can't give off any more heat. This is one of the laws of ***thermodynamics***—the study of heat and how other energy turns into heat.

The first law of thermodynamics states that ***energy cannot be created or destroyed.***

The second law states that ***heat will always flow from an area of higher temperature to an area of lower temperature.***

The third law states that ***it is impossible to reach the state of absolute zero.***

Measuring Heat

Heat is measured in two ways: ***joules*** and ***calories*** (or ***British thermal units***).

Joules

The amount of heat flowing from or into an object is usually measured in joules. A joule (J) is the kinetic energy possessed by a two-kilogram mass moving at one meter per second.

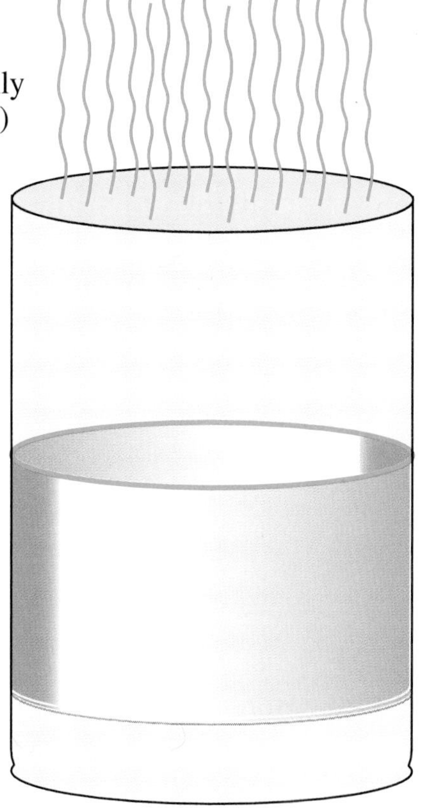

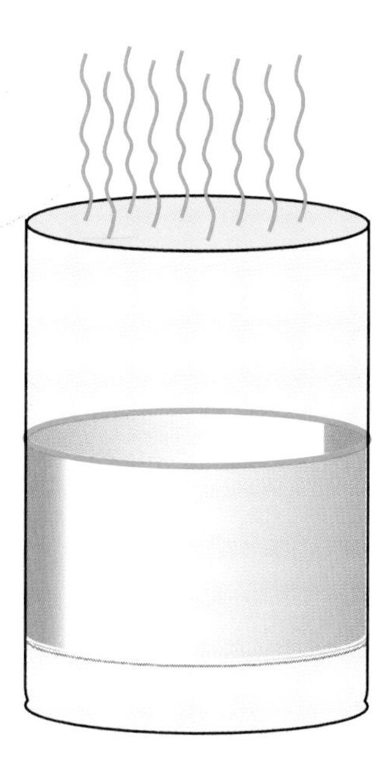

The container with more water at the same temperature has more heat because it has twice the mass/internal energy.

Calories and British Thermal Units

Heat can also be measured by its ability to raise the temperature of water. A calorie is the amount of heat needed to raise the temperature of one gram of water one degree Celsius. A British thermal unit (BTU) is the amount of heat needed to raise the temperature of one pound of water one degree Fahrenheit. A calorie is defined as 4.19 joules, while a kilocalorie used in nutrition is 4,190 joules.

Hot Stuff!

Fahrenheit and centigrade scales are often used to measure body temperature. A healthy body temperature is around 98.6° F or 37°C.

But where does body heat come from?

What Is a Calorie?

Body heat comes primarily from eating food. The heat is measured in ***calories (c)*** and ***kilocalories (C)***, or 1,000 calories. A calorie is a metric measure. It stands for the amount of heat needed to raise the temperature of ***1*** gram of water ***1*** degree centigrade. By measuring the calories in the foods we eat, we can tell how much heat we can generate, or how many Kilocalories we can "burn" in daily activities (Kilocalories are often called Calories, with a capital ***C***).

Calories and Your Body Weight

On days when you eat fewer calories than the number of calories you burn doing activities, you will burn calories stored in the fat and muscles in your body. On days when you eat more calories than you burn through activities, your body will store the excess calories in the form of fat. So, when you eat too little, your body burns up fat. You might even lose weight. When you eat too much, your body stores calories as fat. You might even gain weight.

apple	117 Calories
hot dog	170 Calories
slice of pizza	185 Calories
banana	100 Calories
chocolate (oz.)	155 Calories
carrot	25 Calories
spinach (cup)	23 Calories

Heat and Matter

Heat causes physical and chemical changes in matter (see p. 100).

Physical Changes

Heat changes matter from solid to liquid to gas. These changes are physical changes because the atoms in the substances aren't changed.

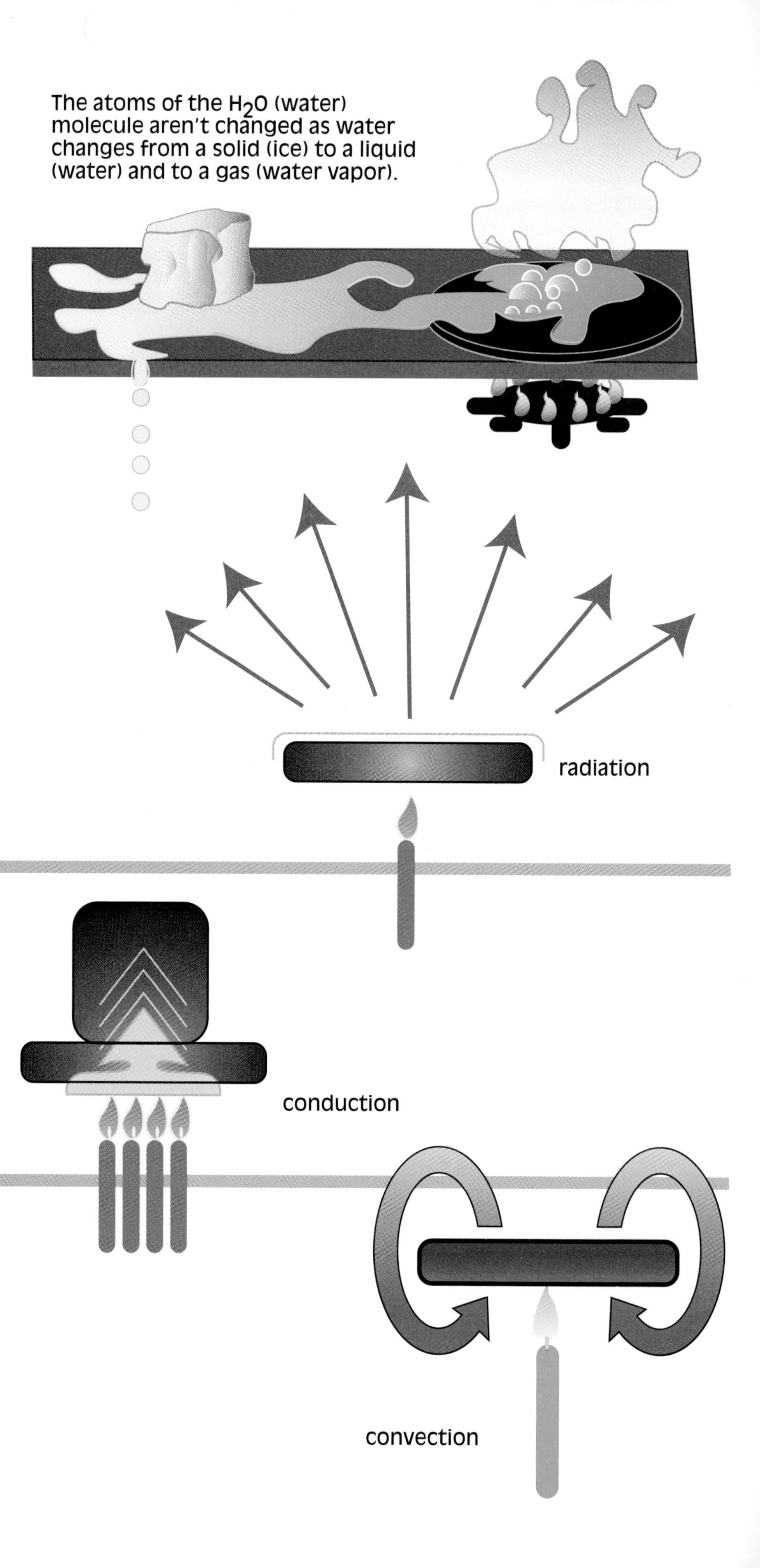
The atoms of the H_2O (water) molecule aren't changed as water changes from a solid (ice) to a liquid (water) and to a gas (water vapor).
radiation
conduction
convection

How Heat Flows

Heat flows from warmer to cooler bodies in three ways: ***radiation, conduction***, and ***convection.***

Radiation

Hot objects give off infrared light in waves that can travel through empty space and air. When this infrared radiation is absorbed by a body, it is changed to heat.

Conduction

Sometimes heat moves through one object into another. This is called ***conduction.*** If you leave a metal spoon in a cooking pot on the stove, the spoon heats up, too. The molecules in the hot food bump into the atoms in the spoon and make them vibrate faster.

Convection

Hot air is not as dense as cool air, so cooler air pushes hot air upward. Heat can move from an energy source in a circular pattern, called a ***convection current.*** Think about a radiator you may have seen. Hot air rises from the radiator, then cools and sinks. As it sinks, it is heated again by the radiator and rises.

TEAKETTLE TEST

A teakettle on a hot stove illustrates the three ways heat travels.

1. **Radiation.**
   The hot teakettle gives off heat.
2. **Convection.**
   Hot water from the bottom of the kettle rises, cools, and sinks, then heats again.
3. **Conduction.**
   Heat from the stove flows from the stove to the metal teakettle.

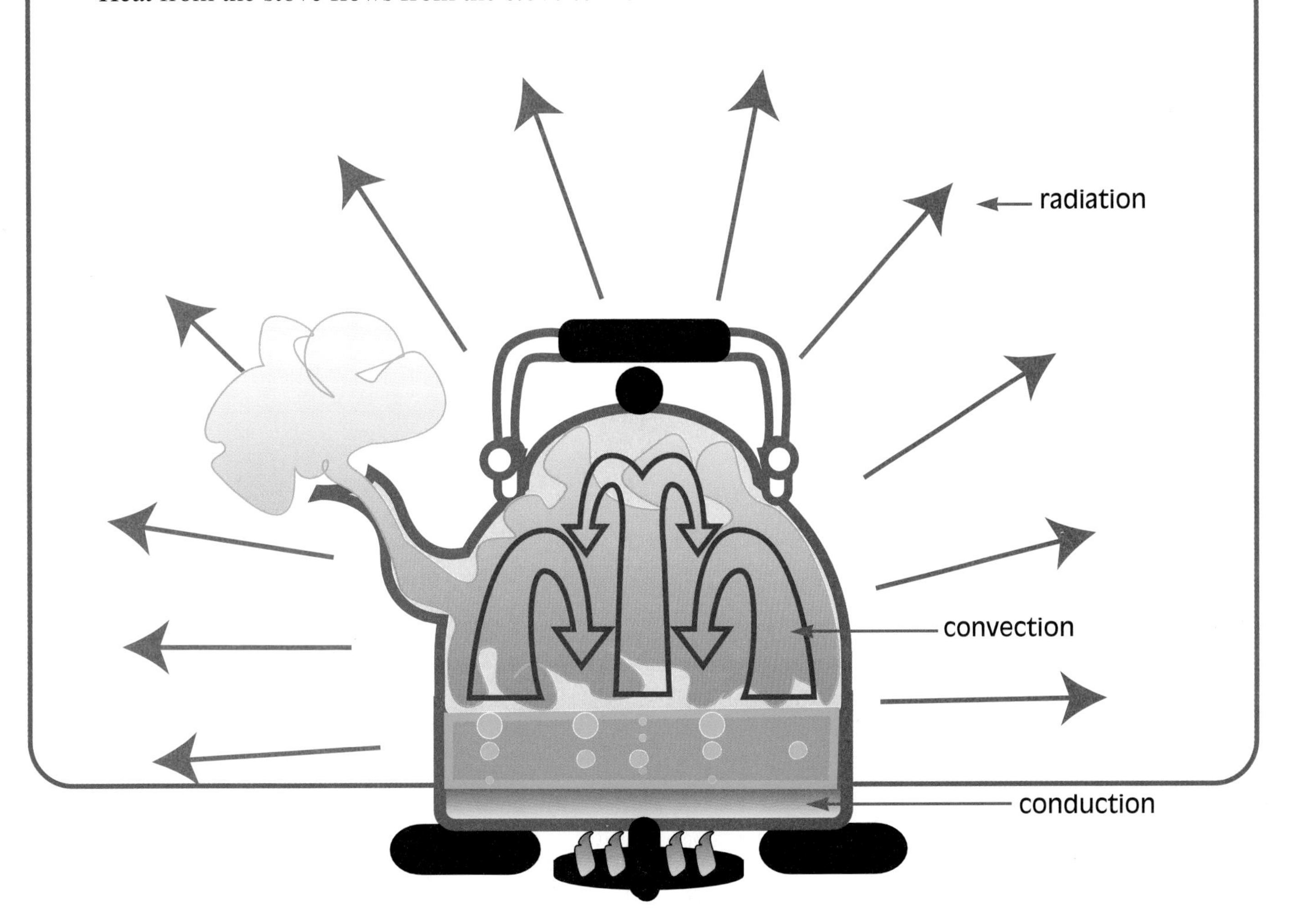

5 Light

Light is energy that acts on our eyes. Without light, we could not see.

Light ***radiates,*** or travels in straight lines in all directions from its source. It consists of tiny particles called ***photons***. Anything that gives off light is called a ***light source***. The sun is our most important natural light source. During hours of darkness, artificial sources, such as electric lights, flashlights, and candles, bring light to our homes, offices, and roads.

Reflection

Light travels only in straight lines. We see objects around us because they ***reflect,*** or bounce, light into our eyes. Light surfaces reflect more light than dark surfaces.

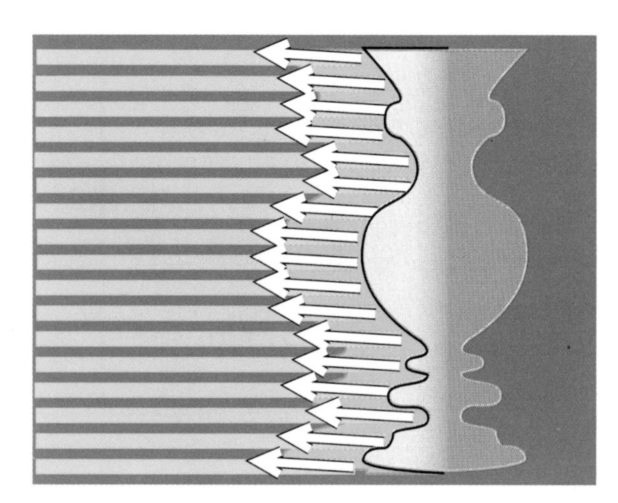

Light travels in straight lines. It is reflected in straight lines, too.

The Law of Reflection

Light that travels through empty space radiates in perfectly straight lines called ***rays*** of light. One way to change the direction of the rays is reflection. A ***mirror***, which can be any shiny surface, is the best tool for reflecting light. With a flat-faced mirror, called a ***plane mirror***, light is reflected at an angle equal to the angle at which the ray hit the mirror (angle of incidence). This law of reflection can be written as a mathematical formula:

The angle of incidence (I) is equal to the angle of reflection (R).

< I = < R

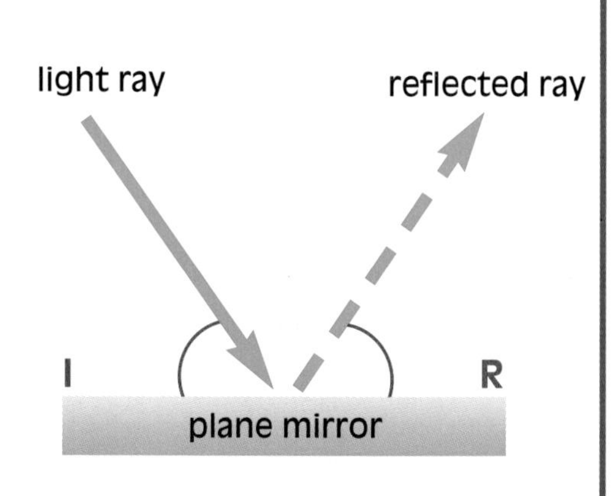

Refraction

Light waves travel faster through air than through water, glass, or other substances. When a light ray passes at an oblique angle from one material to another, its path typically appears to blend. This is called ***refraction.***

Look at a straw in a glass of water. The straw seems to bend at the surface of the water due to *refraction*.

Oasis in the Desert

It's Only a Mirage!

Differences in air temperature and atmospheric pressure can do some funny things to light rays. Pavement on a hot summer day sometimes appears to be wet. But the sidewalk is as dry as a bone. What you're seeing is light refracted by hot air rising up from the pavement. This effect is called a ***mirage.***

Speed of Light

The Danish astronomer Ole Roemer proved in the 1670s that light takes time to travel from one place to another. Over the years, improved instruments have helped scientists calculate the speed of light more accurately, at about 186,282 miles per second (299,792 kilometers per second).

Spectrum

Sunlight seems to be white or colorless. But sunlight is really a combination of colors. The colors we see are reflections of the light that is not absorbed by objects. The different colors of light have different ***wavelengths.*** The range of colors is called the ***spectrum.*** You can see the spectrum in rainbows and when light is ***refracted*** through a prism.

A prism *is a transparent solid with more than two plane faces that are not parallel. When light passes through a prism, it refracts—or bends—often resulting in a visible spectrum.*

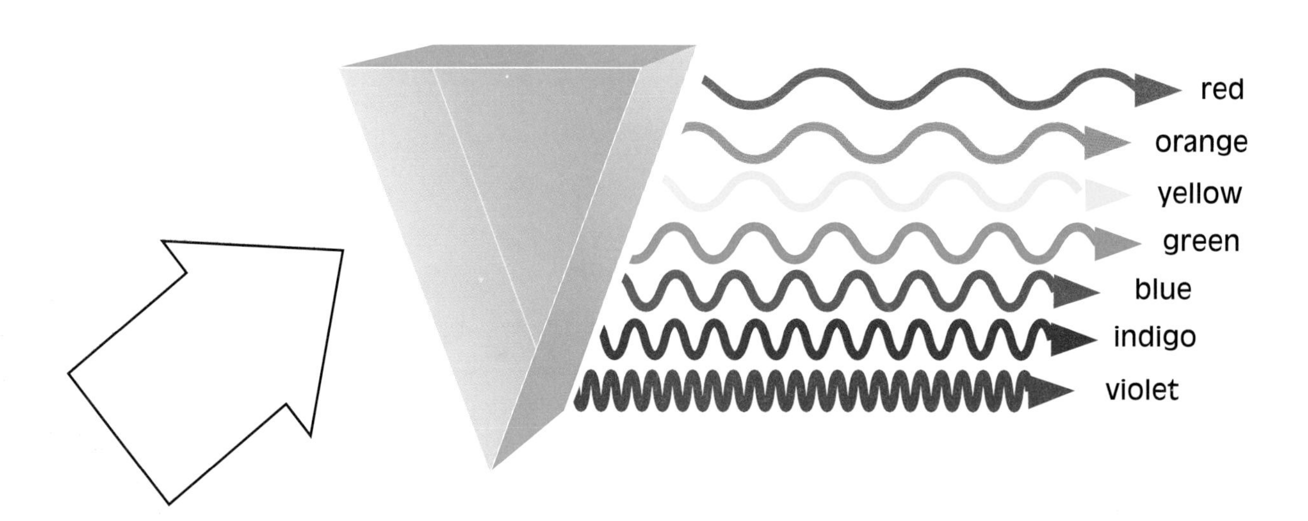

White is the presence of all colors. Black is the absence of color.

The study of light and how it behaves is called optics.

Sound

Sound is anything you hear. Sound is really air ***vibrating***, or moving back and forth quickly. Pluck the string of a guitar or a violin. The string starts vibrating and causes the air around it to vibrate, too. The vibrations travel through the air to your eardrum—and you hear sound.

How Sound Travels

Sound travels from a vibrating object through ***compression waves.*** The more the waves are ***compressed***, or squeezed together, the louder the sound. The loudness of sound is called ***amplitude***.

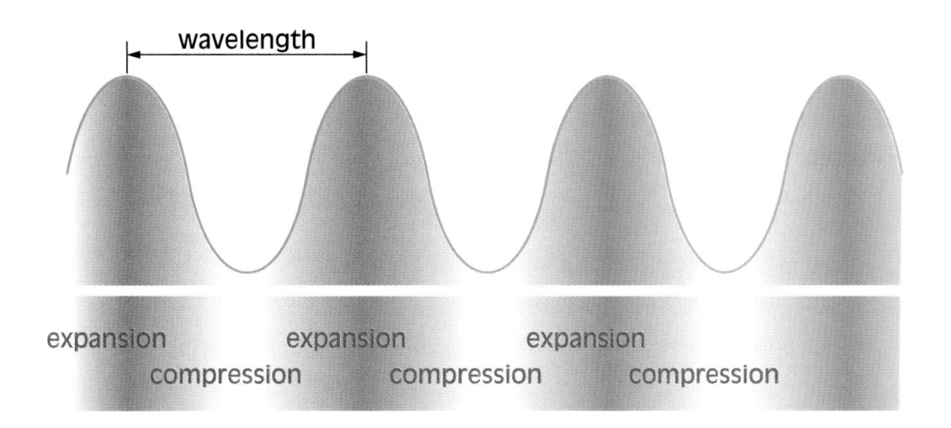

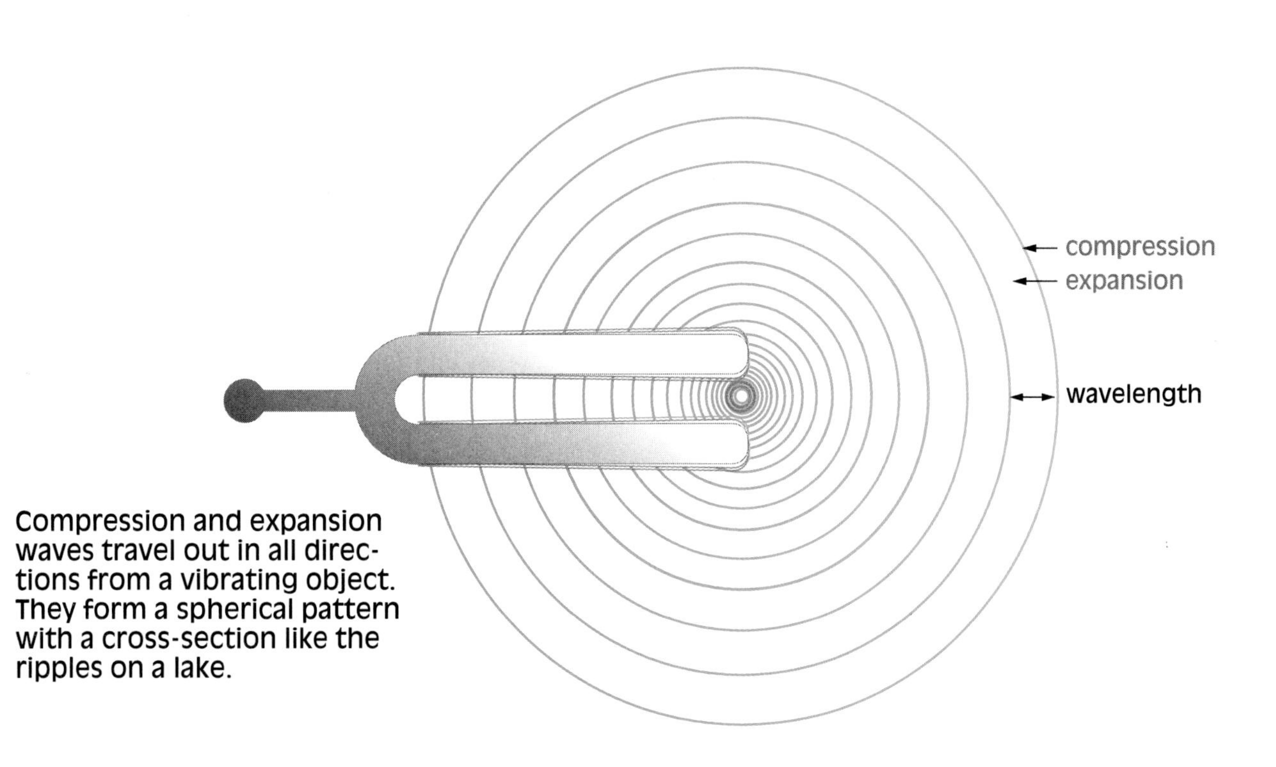

Compression and expansion waves travel out in all directions from a vibrating object. They form a spherical pattern with a cross-section like the ripples on a lake.

Frequency and Pitch

Sound is also described by its ***frequency,*** or the number of its vibrations per second. Frequency is measured in ***hertz*** (Hz), a unit named for German scientist Heinrich Hertz. One hertz means one vibration per second. A doorbell that vibrates 200 times per second has a frequency of 200 Hz. We hear frequency as pitch—***low***, ***medium***, or ***high***—depending on the number of vibrations per second. A frequency of 50 is low, 5,000 is high.

The Speed of Sound

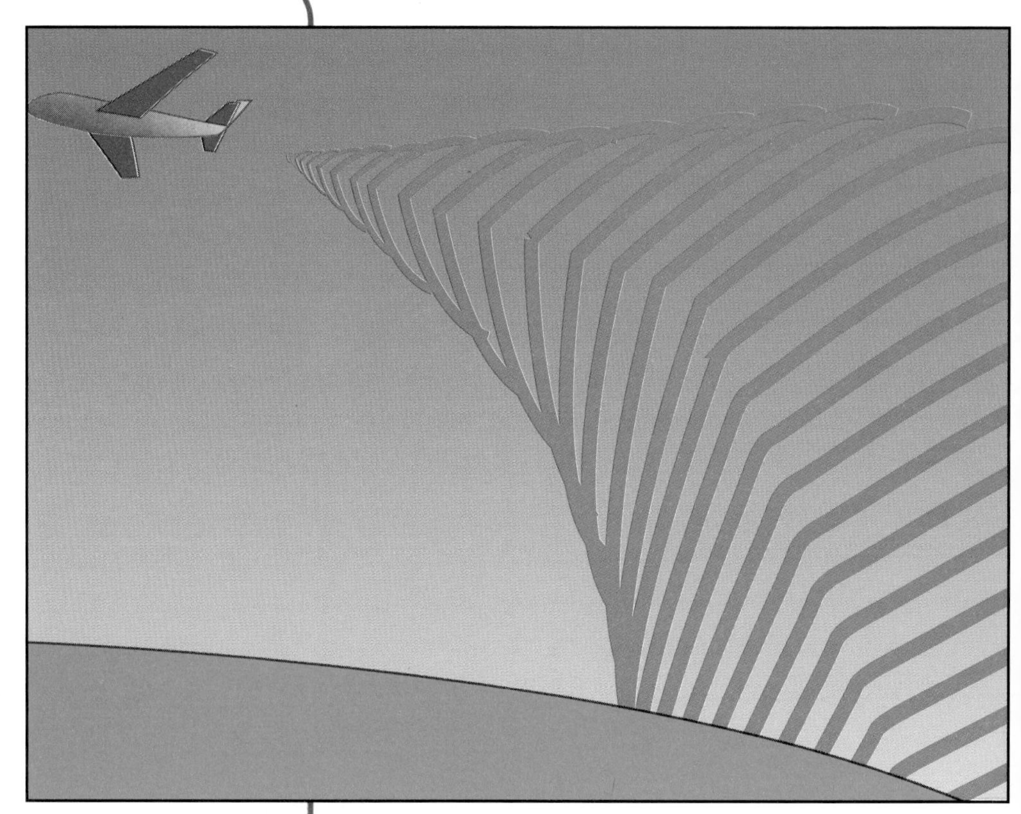

Sound travels about 1,100 feet (335 meters) per second or 740 miles per hour. It travels a little faster on hot days than on cold days. Sound travels four times faster in water, and even faster still through solids than through air.

Sound doesn't travel nearly as fast as light (see Speed of Light, p. 116). Take a close look at the marching bands the next time you see a parade. The marching band members closest to the drummers are slightly out of step from the band members farthest away from the drummers. This is because the faraway band members hear the drumbeats a tiny bit later.

Many airplanes travel faster than sound. When a jet exceeds the speed of sound—740 miles per hour—it breaks the ***sound barrier*** and causes a ***sonic boom.***

Measuring Sound

Sound is measured by its ***intensity***, or loudness. Intensity is called ***sound level,*** and it is measured in units called ***decibels***.

The study of the way sound behaves is called **acoustics.**

HOW LOUD IS IT?

Noise is measured in decibels. The chart shows how some sounds measure up.

160
150
140
130
120
110
100
90
80
70
60
50
40
30
20
10
0

decibel

SOUND EFFECTS

Sound waves don't always reach your eardrums directly from their source. Sometimes sound waves are changed or interrupted. The results can be pretty strange.

I smell a rat.

Is Mel your cat?

Echoes

Echoes are sound waves that are heard separately from waves that travel directly from a source to your ear. You hear a direct wave and then a reflected wave, or echo.

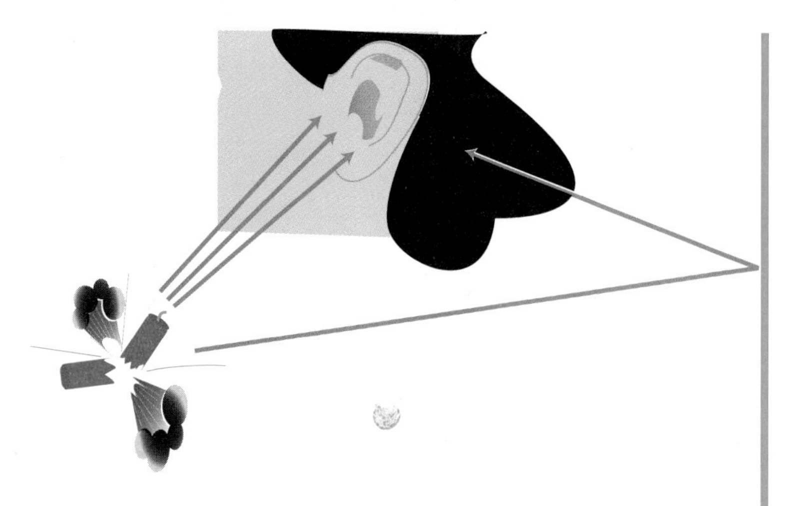

Sound waves travel from an exploding firecracker.

Some sound waves travel directly to your ear.

Other sound waves travel to surfaces that bounce the sound back to the ear. These sound waves arrive at the ear a sufficiently long time *after* the waves that travel directly so that you hear them as a separate sound.

Reverberation

Reverberation is sound that is reflected in several directions between source and ear. If it arrives at the ear late enough to be perceived as a separate sound, it is an echo.

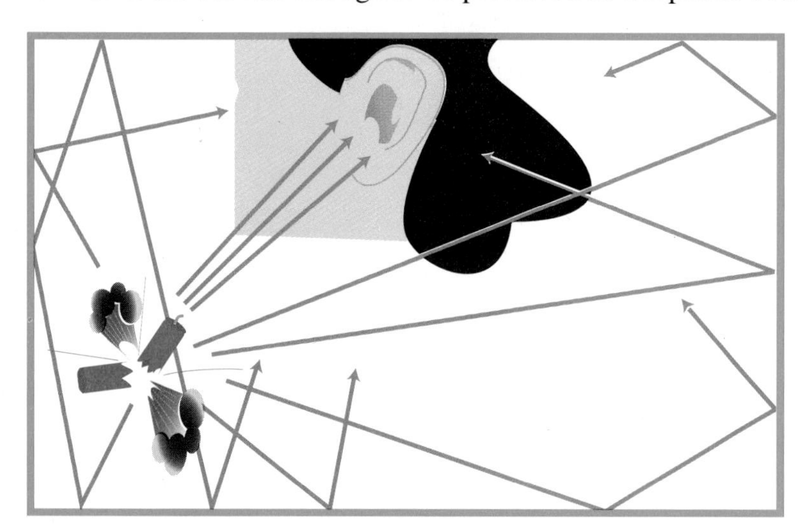

When sound waves are bounced back to the ear from several directions, reverberation is heard.

Ultrasounds: Sounds that Can't Be Heard

Some animals can hear sounds of higher or lower frequency than humans can hear. ***Ultrasonic*** sounds have frequencies higher than about 20,000 Hz—beyond the range of human ears. Ultrasonic wave beams are used in ***sonar*** machines that locate underwater life and objects. A sonar machine called a ***fathometer*** is used to measure the depth of the ocean. Ultrasonic waves are also used to make liquids vibrate quickly. This is useful for stirring paints and homogenizing milk.

Ultrasonics—or ultrasound—is used by doctors to detect brain injuries and blood flow and to monitor the health of unborn babies. Dentists use ultrasound to remove plaque and tartar from teeth.

Electricity and Magnetism

What we typically call ***electricity*** is energy formed by the flow of ***electrons***, the negatively charged particles of atoms (see p. 97). Because everything is made of atoms, all substances have electrons and ***potential*** electricity. A force is all that's needed to get the electricity moving through a conductor (see Potential Energy, p. 101).

Don't Give Me Static — Electricity, That Is

More than 2,000 years ago, the Greek philosopher Thales noticed something special about amber, a brownish resin from pine trees. When he rubbed a piece of amber against woolen cloth, it attracted bits of material. Thales believed amber possessed a special force.

We can conduct Thales's experiment today with a variety of different things, including plastic combs and balloons. Run a comb through your hair on a dry day. Your hair will follow the comb. You can even make your hair stand up straight! Or rub a balloon against your clothes. Then stick the balloon to the wall and let go. It stays on the wall like magic. We know combs and balloons have no more special force than Thales's amber. We know what Thales experienced was ***static electricity.***

Static electricity can be harmless, as you can see from the experiments above. It is the force that gives you a shock when you touch a doorknob after scraping your feet across the rug.

Static electricity can be very powerful, too. Lightning is static electricity. Until the 1700s, people didn't believe that lightning was electrical. Then American statesman and inventor Ben Franklin conducted his famous experiment with a key and a kite. Ben Franklin was lucky he survived—don't ever try his experiment yourself. It could kill you.

Electrical Current

Two Italian scientists—Luigi Galvani and Alessandro Volta—made important discoveries about electricity in the late 1700s. Galvani hung a dead frog on an iron hook and touched the frog with a copper wire. The frog moved. Volta proved that the electricity that made the frog jump came from the wires—not the frog. The frog's muscles served as a ***conductor***, a material that allows electricity to pass through it.

A conductor is a material that allows electrons to flow through it. An insulator is a poor conductor.

After Volta discovered that two metals that touch could create electricity, he built a machine called the ***voltaic pile*** or simple dry cell. The machine created a ***current***, or steady flow, of electricity. Then, a few years later, the English scientist Sir Humphry Davy stacked several voltaic piles and created the first powerful ***battery***.

Circuits

Electrical currents often follow unbroken paths called ***circuits***. Volta and other scientists created electrical current through the use of circuits. There are two types of currents that run through electrical circuits: ***alternating currents*** and ***direct currents***. You can use either one. Most homes in the United States use alternating currents.

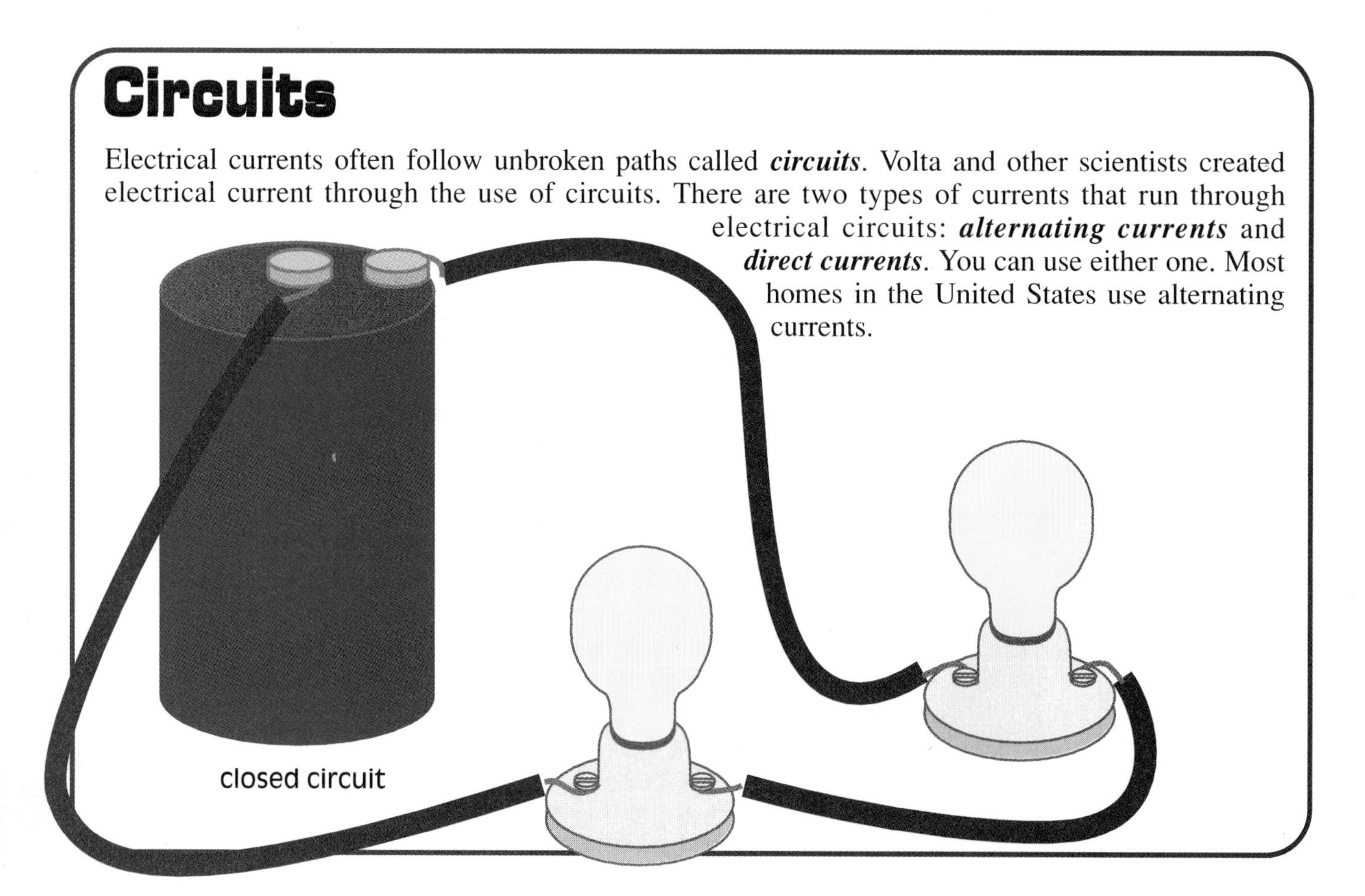

Measuring Electricity

Amperes

The ***current***, or flow, of electricity is measured in ***amperes***, or ***amps***. Amperes measure the number of electrons that flow through one place in one second. The unit ampere is named for Andre Ampere.

Volts

The ***force*** that pushes electrons through an electrical circuit is measured in ***volts***. Volts, or ***voltage***, are named for Alessandro Volta.

Ohms

Electricity travels easily through some materials and less easily through others. The ***resistance*** a material puts against the flow of electricity is measured in ***ohms***. Ohms are named for Georg Ohm.

Watts

The power of an electrical machine is measured in ***watts***. Watts are calculated as voltage times current:

watts = voltage (volts) X current (amps)

Ohm's Law

Georg Ohm was curious about the relationship between current (amps) and force (volts). He noticed in experiments that when the force was cut in half, the current was also cut in half. The resistance—that is, the material that the current and force were measured in—remained the same. And so Ohm came to write down his law.

$$\underset{\text{(in volts)}}{\text{force}} = \underset{\text{(in amperes)}}{\text{current}} \times \underset{\text{(in ohms)}}{\text{resistance}}$$

$$\text{current (in amperes)} = \frac{\text{force (in volts)}}{\text{resistance (in ohms)}}$$

$$\text{resistance (in ohms)} = \frac{\text{force (volts)}}{\text{current (in amperes)}}$$

Magnetism

Magnets are materials that naturally attract iron, nickel, and cobalt. Around 1600, an English physician, William Gilbert, wrote that the earth is one big round magnet and that its poles are magnetically charged. Gilbert demonstrated this with a compass. A compass has a magnetized needle, one end of which points approximately to the earth's north pole. Gilbert was the first to use the term ***magnetic pole***.

It wasn't until 1820 that a Danish scientist, Hans Christian Oersted, noticed that an electrical current flowing through a conductor acted like a magnet. Oersted demonstrated the connection between magnetism and electricity. About the same time, a Frenchman, Andre Ampere, developed the mathematical theories that described electrical currents and magnets.

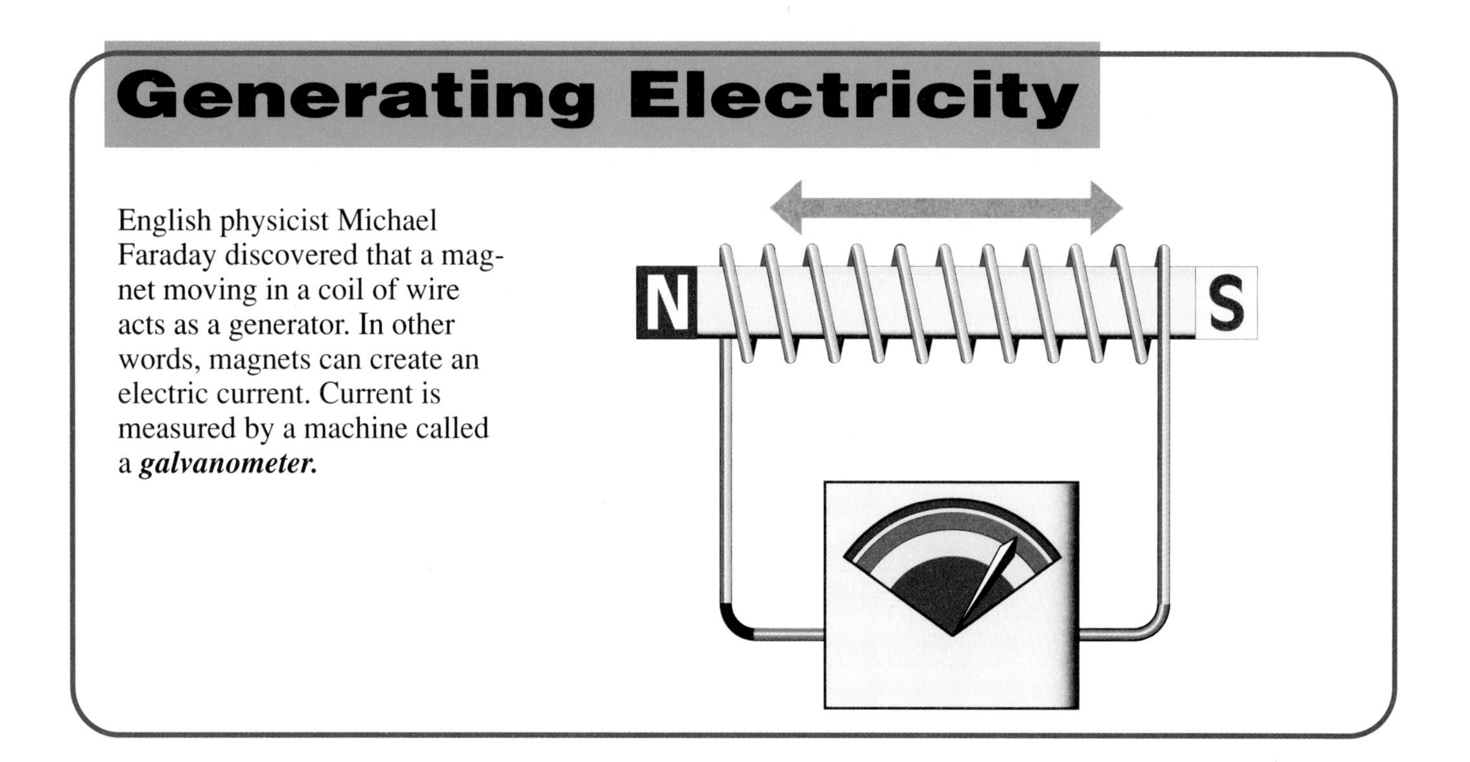

Generating Electricity

English physicist Michael Faraday discovered that a magnet moving in a coil of wire acts as a generator. In other words, magnets can create an electric current. Current is measured by a machine called a ***galvanometer.***

APPENDIX

Reporting on an Experiment

When your teacher tells you to write up an experiment, you can follow the form used by scientists everywhere. There are three parts: the ***hypothesis***, the ***experiment***, and the ***conclusion***.

1 THE HYPOTHESIS

The statement that your experiment was designed to test. You should state your hypothesis simply in a sentence followed by a colon. For example,

Hypothesis: Salt lowers the melting point of water and can be used effectively to melt ice.

2 THE EXPERIMENT has three parts of its own:

Materials: salt, ice, two dishes. Do this experiment in the freezer compartment of a refrigerator (whose temperature is, say, 2°C).

Procedure: Number the steps you perform. For example,

1. I put identical portions of ice in the two dishes.

2. I added salt to the ice in one of the dishes.

3. I put them both in the freezer and left them there __ minutes.

Results: The ice melted in the dish with salt.

3 THE CONCLUSION

The salt caused the ice to melt. This experiment proves my hypothesis that salt melts ice.

INDEX

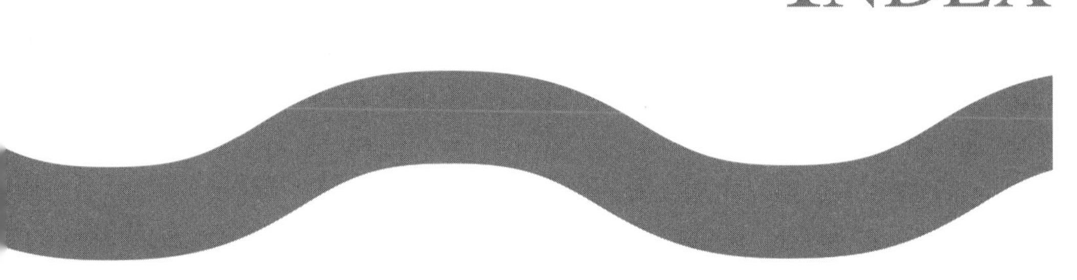

a

b

c

d

e

f

h

i

j

k

l

m

n

p

q

r

S

t

v